# CONTINUOUS EMISSION MONITORING

# CONTINUOUS EMISSION MONITORING

James A. Jahnke, Ph.D.

 VAN NOSTRAND REINHOLD
New York

Printed in the United States of America
For more information, contact:

Van Nostrand Reinhold
115 Fifth Avenue
New York, NY 10003

Chapman & Hall GmbH
Pappelallee 3
69469 Weinheim
Germany

Chapman & Hall
2-6 Boundary Row
London
SE1 8HN
United Kingdom

International Thomson Publishing Asia
221 Henderson Road #05-10
Henderson Building
Singapore 0315

Thomas Nelson Australia
102 Dodds Street
South Melbourne, 3205
Victoria, Australia

International Thomson Publishing Japan
Hirakawacho Kyowa Building, 3F
2-2-1 Hirakawacho
Chiyoda-ku, 102 Tokyo
Japan

Nelson Canada
1120 Birchmount Road
Scarborough, Ontario
Canada M1K 5G4

International Thomson Editores
Campos Eliseos 385, Piso 7
Col. Polanco
11560 Mexico D.F. Mexico

3  4  5  6  7  8  9  10  EDW  99  98  97  96  95

**Library of Congress Cataloging-in-Publication Data**
Jahnke, J. A. (James A.)
      Continuous emission monitoring/James A. Jahnke.
          p.   cm.
      Includes bibliographical references and index.
      ISBN 0-442-00724-8
      1. Continuous emission monitoring.   I. Title
    TD890.J34   1992                      92-27824
    628.5'3'0287—dc20                     CIP

**To Gloria**

# Contents

# Preface

The techniques of continuous emission monitoring (CEM) are widely used today in integrated systems to measure gases and particulate matter emitted from stationary sources of air pollutants. Stationary sources such as coal-fired power plants, municipal waste incinerators, Kraft pulp mills, and industrial process plants are today being required to use CEM systems to provide a continuous record of air pollution control equipment performance and to determine compliance with emission standards.

Continuous emission monitoring constitutes the sum of activities involved in determining and reporting the emission levels of pollutant gases. Instruments have been designed specifically for the measurement of stack gas emissions; however, instruments intended for other applications, such as ambient air monitoring or laboratory analysis, have also been applied.

Environmental regulations have been the driving factor in CEM development and application, both in the United States and in Europe. These regulations have been technology-forcing, defining the performance expected of installed instrumentation. Requirements to monitor emissions at lower and lower concentrations without sacrificing accuracy, precision, or system availability have led to significant improvements in system design. These regulatory requirements have shaped the CEM market and will continue to do so in the future.

This book examines the interplay of science and regulation as it affects the design and certification of continuous emission monitoring systems. It reviews CEM techniques currently used, as well as the regulatory programs that require their installation and continuing performance. This book is intended to be comprehensive in scope, to meet the needs of both the plant environmental engineer applying CEM systems and the control agency's personnel incorporating CEM systems in the regulatory programs.

James A. Jahnke, Ph.D.

# 1

# Introduction to CEM Systems

In the early 1970s a need was recognized for a better way to monitor stack emissions than by conducting manual stack tests. Inserting a probe into a stack, extracting a sample, and analyzing the sample in a laboratory is a time-consuming process. In addition, source operations may be highly tuned during such testing and may not necessarily be representative of day-to-day performance. Clearly, to monitor plant and air pollution control equipment performance, other measurement techniques are necessary.

## A BRIEF HISTORY

Attempts were made in the 1960s to apply ambient air analyzers and process industry analyzers to the measurement of source emissions. The use of ambient air analyzers was not successful at that time due to the instability of necessary dilution systems. However, process analyzers did prove to be applicable, particularly those employing ultraviolet and infrared photometric techniques. Then, in the late 1960s and early 1970s, successful developments in both German and American instrumentation provided more options. Ambient analyzers were redesigned to measure gases at higher levels and the so-called in-situ analyzers, which can measure gases in the stack without sample extraction, were developed. These methods, in addition to new German optical systems for opacity monitors and the development of luminescence measurement techniques in the United States, provided a technological base from which continuous emission monitoring (CEM) regulations could be established.

The first continuous emission monitoring requirements in the United States were promulgated in 1971. However, the CEM industry did not begin to develop until after October 6, 1975, when the U.S. Environmental Protection Agency (U.S. EPA) established performance specifications for CEM systems and required their installation in a limited number of sources. Since that time, CEM systems have been applied to a wider range of sources, and 15 years of experience have led to the evolution of analyzers that can measure stack emissions with a high level of confidence.

The earliest focus of CEM technology was on the analyzer—the instrument that could do the job of measurement. However, it was soon found that the process of transporting the gas to the analyzer was a source of many problems. These problems were addressed in a number of ways by CEM "systems" vendors. Those who understood the effects of corrosive stack gases on materials and the effects of pressure and temperature on gas transport were able to design and successfully market systems that worked under severe sampling conditions.

A CEM system is actually composed of three subsystems: the sampling interface, the gas analyzers, and the data-acquisition–controller system (Figure 1-1). The sampling interface is a subsystem that either transports or separates the flue gas from the analyzer. CEM systems are usually characterized in terms of the design of this interface. In general, the systems can be classified into three basic groups: extractive systems, in-situ

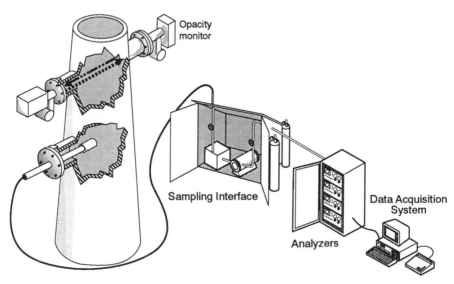

**FIGURE 1-1.**    A continuous emission monitoring (CEM) system.

**TABLE 1-1     Classification of Source Monitoring Systems**

| Extractive Systems | In-Situ Systems | Remote Sensors |
|---|---|---|
| Source-level | Point | Active |
| Dilution | Path | Passive |
| | Single-pass | |
| | Double-pass | |

systems, and remote sensors. An expanded classification is shown in Table 1-1.

In the case of extractive systems, the interface consists of the system designed to extract and condition the gas prior to entering the analyzer. In the case of in-situ systems, the interface is simpler, composed of flanges designed to align or support the monitor and blower systems used to minimize interference from particulate matter. Remote sensing systems in effect have no interface between the stack gases and the sensing instrument, other than the ambient atmosphere.

## EXTRACTIVE SYSTEMS

Extractive gas-monitoring systems were the first to be developed for source measurements. In these systems, gas is extracted from a duct or stack and is transported to analyzers for the measurement of the pollutant concentrations. Many of the early extractive systems first diluted the gas using rotameters and then applied ambient air analyzers for the measurements. However, frequent problems occurred in maintaining stable dilution ratios, so analyzers were subsequently developed to measure the flue gas directly at source-level concentrations. These source-level extractive systems have proven to be quite successful and received their widest application in the 1970s and early 1980s.

Many of the problems associated with the earlier dilution systems have since been eliminated by new techniques, developed in the 1980s. The advent of the "dilution probe" has made dilution systems viable for source measurements. These systems are now relatively easy to construct and exhibit good performance.

The basic problem associated with extractive systems is that they are, indeed, "systems." In order for an instrument to measure gas concentrations, the gas sample must be free of particulate matter. Water vapor usually must be removed and the sample must be cooled to instrument temperature. This requires the use of valves, pumps, chillers, sample tubing, and other components necessary for gas transport and conditioning. These components require maintenance, which must be performed routinely. Failure to perform required maintenance and problems associated with poor system designs have led to frequent probe plugging, corrosion, and leaks. These problems discouraged many early users of these systems.

## IN-SITU SYSTEMS

The difficulties associated with extractive systems led to the idea of measuring the flue gases as they exist in the stack or duct, without conditioning. This idea was realized in the development of "in-situ" analyzers, a second generation of source monitoring systems especially designed to avoid the problems inherent in extracting gases. These systems consist primarily of an analyzer that employs some type of sensor to measure the gas directly in the stack or that projects a light through the stack to make the measurement.

There are two classifications of in-situ analyzers: point and path. Point analyzers consist of an electrochemical or electro-optical sensor mounted on the end of a probe that is inserted into the stack. The in-stack, point measurement is usually made by the sensor over a distance of only a few centimeters. Path analyzers, on the other hand, measure along a path across the width of the duct or diameter of the stack. In these "cross-stack" analyzers, light is transmitted through the gas, and the interaction of the light with the flue gas is used to obtain a quantitative value of pollutant concentrations. In the case of single-pass instruments, light is transmitted from a unit on one side of the stack to a detector on the other side, making only one pass through the stack. In a double-pass system, the light is reflected from a mirror on the opposite side, doubles back on itself, and is detected back at the "transceiver."

In-situ analyzers are used to measure flue gas opacity and the concentrations of pollutant and combustion gases. Opacity monitors (transmissometers) can be either single-pass or double-pass systems and use visible light to make the opacity measurement. Instruments are also being developed to monitor particulate mass concentration as an emission parameter.

## REMOTE SENSORS

A third generation of source emission monitors has developed from the technology of the U.S. space program. These monitors are remote sensors that can detect emission concentrations merely by projecting light up to the stack (active systems) or by sensing the light radiating from the "hot" molecules emitted from the stack (passive systems). Already, the U.S.

(EPA) has developed a reference method (method 9A) for monitoring opacity, using a laser light detection and ranging (LIDAR) technique. However, neither reference methods nor performance specifications have been standardized for remote sensing systems for gases.

Due to an inherent problem in defining the length of the measurement path in the plume, the accuracy of gas concentration data is poorer than that obtained by the extractive or in-situ techniques. Also, because the regulatory applicability of remote measurements of pollutant gas to stationary sources has not been established as yet, the development of these systems has been slowed and will not be discussed further.

## ANALYTICAL TECHNIQUES USED IN CEM SYSTEM INSTRUMENTATION

The analytical techniques used in extractive and in-situ CEM systems encompass a wide range of chemical and physical methods. These vary from chemical methods using simple electrochemical cells to advanced electro-optical techniques such as gas filter correlation spectroscopy and Fourier transform infrared spectroscopy. Table 1-2 gives a summary of the

**TABLE 1-2    Analytical Techniques Used in Continuous Emission Monitoring Systems**

| | In-Situ Systems | |
| --- | --- | --- |
| Extractive Systems | Gases | Particulate Matter |
| Absorption spectroscopy: | Point: | Point: |
| Spectrophotometry | Absorption spectroscopy | Light backscattering |
| Differential absorption |   Second-derivative spectroscopy | Ion charge transfer |
| Gas filter correlation | Electroanalytical | Nuclear radiation |
| Fourier transform infrared |   Polarography |   attenuation |
| |   Electrocatalysis | |
| Luminescence methods: | Path: | Path (opacity): |
| Fluorescence ($SO_2$) | Absorption spectroscopy | Light scattering |
| Chemiluminescence ($NO_x$) |   Differential absorption |   and absorption |
| Flame photometry ($SO_2$) |   Gas filter correlation | |
| Electroanalytical methods: | | |
| Polarography | | |
| Potentiometry | | |
| Electrocatalysis ($O_2$) | | |
| Paramagnetic methods ($O_2$) | | |

*Note:* The name of a gas in parentheses following a method indicates that the technique is currently commercially applied only to that gas.

analytical techniques used in currently marketed CEM systems. Techniques used for laboratory analysis, as well as techniques developed specifically for emissions monitoring, have been incorporated into commercially marketed systems. New analyzers have been developed using established electro-optical methods, updated with solid-state circuitry and microprocessors.

In the succeeding chapters, this book discusses the details of both extractive and in-situ systems—their advantages and disadvantages and their limits of application. The sampling interface is of particular importance in extractive system design and is treated separately in Chapter 3. For in-situ system design, the analyzer type is most important. In-situ monitors for measuring gases are discussed in Chapter 6, and monitors designed for measuring flue-gas opacity are discussed in Chapter 7.

## DATA ACQUISITION SYSTEMS

An important component of any CEM system is the data acquisition system (DAS). CEM data acquisition systems have evolved along with advances in the computer industry. In the 1970s, strip chart recorders were initially deemed adequate, but microprocessors and personal computers were rapidly applied to minimize the tedious review of the continuous data. Increasingly stringent agency reporting requirements have also necessitated the application of automated data handling systems.

The data acquisition system has been used frequently to serve two functions: (1) to control the automatic operations of the system, such as daily calibration, and (2) to process system data. Increasingly, control functions are being separated from the DAS by using programmable logic controllers (PLCs) for system control. These subsystems then report status conditions and input data to the DAS. The choice between the two methods can be one of the distinguishing features between CEM system vendors.

CEM data acquisition system with graphics capability, multitasking functions, and telemetry are being increasingly demanded by CEM system users. Implementation of the U.S. Clean Air Act Amendments of 1990 will require more in terms of electronic reporting from CEM systems, and it is expected that CEM systems will undergo significant development in this area.

## THE ROLE OF QUALITY ASSURANCE

In the 1970s, industry frequently presented quite valid arguments that the performance of continuous monitoring systems was questionable. Two basic principles of CEM technology were soon learned:

1. There is no one "best" system for all applications.
2. A CEM system must be maintained if it is to operate.

During this period, aggressive CEM system vendors frequently sold their systems to anyone who could be convinced to buy one. This resulted in misapplications of both in-situ and extractive systems. The resulting poor performance led to unfortunate perceptions about the reliability of the technology and to the bankruptcy and absorption of several companies. From this experience, formal procedures for specifying and evaluating CEM systems have been developed and are being used increasingly by companies planning major CEM system purchases.

However, errors in application were not the only reason for poor CEM system performance. It was often assumed that after a CEM system was installed it could generate data as routinely as a thermocouple or pressure gauge. It was not realized that routine maintenance programs were necessary for the continuing operation of the extractive system plumbing and electro-optical systems. Although this necessity was well known in Germany, where responsibility for system operation typically is assigned to trained technicians, an awareness of this need did not develop in the United States until the early 1980s. A CEM specialty conference of the Air Pollution Control Association held in Denver in 1981 pointed out the need for established and effective CEM system quality assurance (QA) programs. By the time of a subsequent conference held in Baltimore in 1985, the U.S. Environmental Protection Agency had proposed CEM system quality assurance requirements and many companies were reporting the success of their own QA programs in improving CEM system performance.

Like an automobile, where the oil must be changed and the tires rotated, a CEM system requires routine checks and replacements. QA programs incorporating daily and weekly checks, periodic audits, and preventive maintenance procedures have been found to be the key to continuing CEM system operation. Systems with such programs today show better than 95% data availability.

## APPLICATION

In the United States, CEM systems were originally required for monitoring the effectiveness of air pollution control equipment in removing pollutants from flue gases. This application has been extended to determine source compliance with emissions standards and, today, to an increasing role in addressing public concerns over stack emissions.

Although emission monitoring systems have been applied principally to satisfying such regulatory requirements, the systems can prove beneficial in plant operations. They can be used to improve process efficiency and to decrease control equipment operating costs. In addition, they can be used to gather design information and to determine maintenance needs. CEM system data can also be used proactively by plant and corporate management by providing a base of information on compliance status or for consideration in legal issues. Such a data base can also be used to provide assurance to the public that emissions are being monitored to address environmental concerns.

Agency uses of CEM systems and CEM system data are, of course, more extensive. Whether used to track pollutant control system performance through "excess emission reports," submitted quarterly, or directly for enforcement purposes, local, state, and federal agencies are depending increasingly on CEM data. CEM systems provide the basis for the acid rain control program mandated in the 1990 Clean Air Act Amendments. Here, CEM systems will determine the "allowances" (the number of tons per year of $SO_2$ emissions) that will be traded between the electrical utilities.

## SUMMARY

The technology of continuous emissions monitoring has not been static. Requirements for the installation of continuous monitors on new categories of stationary sources, the demands for increased system availability, and advances in data handling and reporting have led to more-sophisticated systems with better reliability. CEM systems have advanced considerably over the past 15 years, with improved sampling techniques, analyzers, and data processing systems being integrated to meet the challenges posed by new requirements. Also, by implementing CEM system quality assurance programs and by properly managing the monitoring program, high system availability can be achieved. This high availability is a necessity today, where inaccurate or missing data can incur both regulatory and economic penalties.

### Bibliography

Air Pollution Control Association. 1981. *Proceedings—Specialty Conference on Continuous Emission Monitoring: Design, Operation and Experience.* Air Pollution Control Association, Pittsburgh.

Air Pollution Control Association. 1985. *Transactions—Continuous Emission Monitoring: Advances and Issues*. Air Pollution Control Association, Pittsburgh.

Air Pollution Control Association. 1990. *Proceedings—Continuous Emission Monitoring: Present and Future Applications*. Air Pollution Control Association, Pittsburgh.

Electric Power Research Institute. 1983. Continuous emissions monitoring guidelines. Report No. EPRI CS-3723. Palo Alto, CA.

Electric Power Research Institute. 1988. Continuous emissions monitoring guidelines update. Report No. EPRI CS-5998. Palo Alto, CA.

Jahnke, J. A. 1984. *Transmissometer Systems—Operation and Maintenance, An Advanced Course*. EPA 450/2-84-004.

Jahnke, J. A., and Aldina, G. J. 1979. *Continuous Air Pollution Source Monitoring Systems—Handbook*. EPA 625/6-79-005.

Lillis, J. E., and Schueneman, J. J. 1975. Continuous emission monitoring: Objectives and requirements. *J. Air Pollut. Control Assoc.* 25(8):804–809.

Kiser, J. V. L. 1989. Continuous emissions monitoring: A primer. *Waste Age* May 1988.

Makansi, J. 1989. Move toward process control for CEM natural but slow. *Power* August 1989.

Willard, H. H., Merritt, L. L., and Dean, J. A. 1987. *Instrumental Methods of Analysis*. Van Nostrand, New York.

# 2

# Implementing Regulations

Continuous emission monitoring technology has developed principally by regulatory demand. Where there is a clear economic benefit, flue-gas monitoring systems are readily applied. By the 1960s, instrumentation had been applied to monitoring product loss in the process industries, but it was not until environmental control agencies began requiring the installation of monitoring systems that the CEM industry began to develop.

This regulatory development began in both the United States and the Federal Republic of Germany (FRG). Continuous monitoring requirements were first promulgated for fossil-fuel-fired steam generators in the United States in December 1971. In 1974, Germany passed the Federal Emission Control Law, which incorporated continuous monitoring requirements. Also in 1974, pollutant emission limits and further monitoring requirements were published in the German *Technical Instructions on Air Quality Control* (TA-Luft).

However, intensive development of monitors did not begin until 1975. The U.S. EPA published "performance specification procedures" for continuous emission monitors, and the German Federal Ministry of the Interior (BMI) published CEM "suitability testing guidelines." Therefore, in 1975, instrument manufacturers and sources required to install CEM systems now had criteria by which to design and evaluate monitoring systems. Instruments were developed in both Germany and the United States, using well-known spectroscopic and electrooptic techniques. In addition, in-situ analyzers and a number of innovative electrochemical techniques began their development in the United States. Today, both German and U.S. instrumentation dominate the CEM system market. This market is growing through the increasing recognition by environmental

10

control agencies of the practical applicability of CEM systems in regulatory programs.

CEM technology has now matured and is accepted as a viable regulatory tool. CEM requirements were specified by the U.S. Congress in the 1990 Clean Air Act Amendments, and member states of the European Community will soon be adopting CEM requirements in their control programs. Continuous monitoring instrumentation is now used for a number of regulatory purposes, but, significantly, it is being used increasingly to enforce source compliance with emissions standards. The technology thus provides a valuable tool in national environmental control programs for the protection and improvement of air quality.

## UNITED STATES FEDERAL PROGRAMS

National environmental control programs take many forms. They are generally directed from a central agency that allows states, provinces, or districts some level of local control. The central agency generally has the resources to conduct studies and research programs needed to provide a technical basis for emissions standards and test methods. These standards and methods are then incorporated (and in some cases modified) by the state and district agencies in their control programs.

In the United States, emissions standards and monitoring requirements for stationary sources are drafted by offices of the federal EPA. The most important of these are the Office of Air Quality Planning and Standards (OAQPS), the Office of Solid Waste, and the Office of Atmospheric and Indoor Air Programs. These offices prepare background documents, drafts of proposed regulations, and monitoring requirements. After public hearings, comment, and revision, the regulations and requirements are promulgated and adopted into the *U.S. Code of Federal Regulations* (CFR).

The CFR is a multivolume compendium of U.S. regulations for federal government agencies. The CFR is revised only annually, but is supplemented by the *Federal Register*, which is published each government business day and includes regulatory proposals, notices, and discussions of proposed and promulgated regulations.

The U.S. EPA regulations that affect stationary sources are found under Title 40 of the code. Newly constructed sources are required to meet "new source performance standards" (NSPS), which are given in Part 60 of Title 40 (expressed as 40 CFR 60). Each source category (such as electric utilities, municipal incinerators, or cement plants) is assigned a subpart letter (e.g., Subpart Da, Ea, and F, respectively) by which it is referred to in the CFR. Also, regulations for industrial furnaces that incinerate hazardous wastes have been promulgated by the Office of Solid

**TABLE 2-1    Summary of NSPS and Other Continuous Emission Monitoring Requirements (U.S. Government 1991)**

| Source Category[a] | Source Facility | Proposed Date | Promulgated Date | Requirements |
|---|---|---|---|---|
| New Source Performance Standards 40 CFR 60 | | | | |
| Fossil-fuel-fired steam generators (D) | Boilers > 73 MW | 8/17/71 | 12/23/71 | Opacity, $SO_2$, $NO_x$ |
| Electric utility steam generating units[b] (Da) | Boilers > 73 MW | 9/19/78 | 6/11/79 | Opacity, $SO_2$, $NO_x$ |
| Industrial–commercial–institutional steam generating units[c] (Db) | Boilers > 29 MW < 73 MW | 6/19/84 | 11/25/86 | Opacity, $SO_2$, $NO_x$ |
| Small industrial–commercial–institutional steam generating units (Dc) | Boilers > 2.9 MW < 29 MW | 6/9/89 | 9/12/90 | Opacity, $SO_2$ |
| Portland cement plants (F) | Kiln and clinker cooler | 8/17/71 | 12/23/71 12/14/88 (revision) | Opacity |
| Municipal waste combustors (Ea) | Combustor | 12/20/89 | 2/11/91 | Opacity, $SO_2$, $NO_x$, CO |
| $HNO_3$ plants (G) | Process equipment | 8/17/71 | 12/23/71 | $NO_x$ |
| $H_2SO_4$ plants (M) | Process equipment | 8/17/71 | 12/23/71 | $SO_2$ |
| Petroleum refineries (J) | Catalytic cracker | 6/11/73 | 3/8/74 | Opacity, $SO_2$, $O_2$ |
| | Fuel gas combustor | | | $SO_2$ |
| | Claus recovery plants | | | $SO_2$ |
| Primary copper smelters (P) | Roaster–smelter, CU converter | 10/16/74 | 1/15/76 | $SO_2$ |
| | Dryer | | | Opacity |
| Primary zinc smelters (Q) | Sintering machine | 10/16/74 | 1/15/76 | Opacity, $SO_2$ |
| Primary lead smelters (R) | Blast or reverberatory furnace, | 10/16/74 | 1/15/76 | Opacity |
| | Sintering machine | | | Opacity, $SO_2$ |
| Ferroalloy production facilities (Z) | Submerged electric arc furnaces | 10/21/74 | 5/4/76 | Opacity |
| Steel plants (AA) | Electric arc furnaces | 10/21/74 | 9/23/75 | Opacity |
| Steel plants (Aaa) | Electric arc furnaces | 8/7/83 | 10/31/84 | Opacity |
| Kraft pulp mills (BB) | Recovery furnace | 9/24/76 | 2/23/78 | Opacity, TRS[d] |
| | Lime kiln | | | TRS |
| | Digester | | | |
| | Brown stock washer | | | |
| | Evaporator; oxidation and stripper system | | | |
| Glass manufacturing plants (CC) | Glass melting furnaces | 6/15/79 | 10/7/80 | Opacity |
| Lime manufacturing facilities (HH) | Rotary lime kiln | 5/3/77 | 3/7/78 | Opacity |
| Phosphate rock plants (NN) | Dryer and calciner | 9/21/79 | 4/16/82 | Opacity |
| | Grinder | | | Opacity |

| Source Category[a] | Source Facility | Proposed Date | Promulgated Date | Requirements |
|---|---|---|---|---|
| New Source Performance Standards 40 CFR 60 | | | | |
| Flexible vinyl and urethane coating and printing (FFF) | Solvent recovery controls | 1/18/83 | 6/29/84 | VOC[e] |
| Onshore natural gas processing (LLL) | Sweetening units | 1/20/84 | 10/1/85 | Velocity |
| Petroleum refinery wastewater systems (QQQ) | Carbon adsorbers | 5/4/87 | 11/23/89 | VOC |
| Magnetic tape coating facilities (SSS) | Carbon adsorbers | 1/22/86 | 10/3/88 | VOC |
| Polymeric coating of supporting substrates facilities (VVV) | Carbon adsorbers | 4/30/87 | 9/11/89 | VOC |
| Resource Conservation and Recovery Act 40 CFR 264/266 | | | | |
| Hazardous waste incinerators | Incinerators | 12/19/80 | 1/23/81 | CO, velocity |
| Boilers and industrial furnaces burning hazardous wastes (BIF rules) | Boiler, furnace, or kiln | 5/6/87 10/26/89 11/27/90 | 2/21/91 | CO, $O_2$, THC[f] |
| Clean Air Act Amendments of 1990 (40 CFR 75) | | | | |
| Steam and electric generators | Boilers > 75 MW | 12/3/91 | 1/11/93 | Opacity, $SO_2$, $NO_x$, velocity |

[a]Letters in parentheses indicate subpart letter.
[b]Built between 1971 and 1978
[c]Built after 1978
[d]Total reduced sulfur
[e]Volatile organic carbon
[f]Total hydrocarbons

Waste and are given in Part 264 and 266 (40 CFR 264 and 40 CFR 266). Both new and existing electric utilities must also meet the acid rain requirements established in the 1990 U.S. Clean Air Act Amendments. The acid rain rules are prepared by the U.S. EPA Office of Atmospheric and Indoor Air Programs.

The rules prepared by these offices primarily affect newly constructed sources. A *new source* is generally defined as one constructed after the date the rules are first proposed in the *Federal Register*. An *existing source* is a source constructed before that date. Rules for existing sources are developed by the individual states, with federal guidance and approval.

Table 2-1 lists those *source categories* required to monitor concentrations of gaseous pollutants and/or flue-gas opacity. The table also lists the types of *facilities* at the source, on which a CEM system is to be installed. A facility is defined here as an operational unit of the source, such as a boiler or kiln. For example, Subpart BB for Kraft pulp mills may require monitoring of a number of facilities (such as the recovery furnace, lime kiln, and digester) contained within the plant. Sources such as electric utilities or municipal waste combustors may have from one to six boilers or combustors, all of which may require monitoring.

From Table 2-1, it can be seen that in the United States over 18 categories of new sources are now required to install some type of continuous monitoring system. Other sources may be required to monitor process parameters such as pressure drop, temperature, or flow rate. Because the subparts tend to be very complex, they should be referred to for detailed information concerning units of emissions standards, monitoring requirements, reporting requirements, and exceptions.

States may prepare rules for existing sources that are more stringent than those promulgated by the federal agency (Kerstetter 1985). These often incorporate CEM requirements.

## REGULATORY USES FOR CEM DATA

The most obvious impact of continuous data is that more information is available to environmental control agencies. By obtaining such data (or summaries of data), fewer on-site inspections and manual reference method tests may be necessary to enforce emissions standards. Different approaches can be taken by an agency in applying CEM data to meet its goals of maintaining and improving air quality. Properly used, CEM data can expand an agency's enforcement capabilities. The CEM data base can assist in negotiating the installation of control equipment or in requiring process modifications. Most recently, in the acid rain program mandated by the 1990 Clean Air Act Amendments, CEM data will provide the basis for a supply-and-demand trading market for $SO_2$ emissions.

### Using CEM Systems for Reporting Excess Emissions

Federal CEM requirements were originally developed in 1975 to provide a means for sources and agencies to check the operation and performance of NSPS-mandated control equipment such as bag houses and wet scrubbers. The intention was that the source could use the emissions data to track air pollution control system performance and that reports would be submitted quarterly to the agency. These reports, called excess emission reports (EERs), were to include data on exceedences of the emissions standards.

In 1975, new fossil-fuel-fired steam generators (FFFSGs), petroleum refineries, nitric acid plants, and sulfuric acid plants were the first source categories in the United States required to report such data on a quarterly basis. The excess emission reports can be used for the following:

1. to indicate if the source is using good operating and maintenance practices on its process and its control equipment to minimize emissions
2. to provide the agency with data on upset conditions or trend data indicating degradation of air pollution control equipment performance
3. to provide the agency with a continuous record of the source's ability to comply with standards
4. to provide a screening tool in inspection targeting programs
5. to provide the agency with sufficient data to issue a notice of violation (NOV) if the source is not complying with regulations or standards

Under U.S. regulations where CEM data are used for excess emissions reporting, the data usually cannot be used alone as legally enforceable determinants of when emission limits have been exceeded. Instead, the data show that a problem exists and, based upon the severity of the problem, the agency takes further action, which might include conducting the reference method tests for determining compliance with emissions standards. The reference method test data would then determine if the source was complying or not complying with its emissions standards.

Excess emission report (EER) data are not used directly to refer cases to the Department of Justice or to issue an order of noncompliance, unless other regulatory mandates specify the CEM system as the compliance method [U.S. EPA—Office of Air Quality Planning and Standards 1986 (E. Reich)]. Actually, this limitation does not preclude the use of EER data in enforcement cases. Because an agency may review all available information when evaluating the compliance status of a source, the CEM data can constitute part of the record of an enforcement proceedings. Section 113 of the Clean Air Act states:

"Whenever, on the basis of any information available to him, the Administrator finds that any person is in violation of any standard of performance . . . he may bring a civil action in accordance with subsection 113(b)."

The U.S. EPA can use EER data alone as a basis for issuing a "finding of violation" (FOV) or a "notice of violation" (NOV) [U.S. EPA—OAQPS

1986, (Reich)]. If the source does not come into compliance after the FOV or NOV is issued, the EPA must acquire compliance test (reference method) data before it can initiate litigation or issue a "notice of noncompliance" (NON). U.S. EPA Region V has issued NOVs on the basis of the Clean Air Act Section 113 for effectively enforcing source compliance (McCoy 1985).

The advantage of the excess emission approach for source enforcement is that it allows an agency discretion in its enforcement program. Based upon past source performance, agency resources, and so on, an agency may choose to follow up or not to follow up on reported instances of excess emissions. A phone call may suffice to clarify a problem, an NOV may be issued, or, further, a site inspection or source test may become necessary.

### Enforcement Levels of the EER Approach

Determination of source compliance can be made on a number of levels when using the EER approach.

#### Level 1
On the first level, the agency evaluates the source's quarterly EERs, checking the following items:

- Reports of periods and magnitudes of excess emissions
- Nature and cause of each period of excess emissions
- Periods during which the continuous monitoring system was inoperative
- Records of calibration checks, adjustments, and maintenance performed on the monitoring system

Problem areas in source operation and/or control equipment operation should present themselves upon thorough evaluation of the preceding items. These administrative evaluations can save agency labor and expense. The agency can contact the facility for clarification of the problem areas; this might avoid more extensive agency action.

The excess emission reports can be used alone to initiate other levels of activity. For example, in U.S. EPA Region IV, a problem source is considered to be one that is not in compliance with an emission specification greater than 5% of its total monitoring time or one that has monitor downtime greater than 5% of its total operating time. Table 2-2 gives the actual criteria for follow-up actions.

Other programs may mandate further steps before requiring testing. These include conducting either systems audits and/or performance audits.

**TABLE 2-2    EPA Region IV Target Criteria and Follow-Up Actions where CEM Systems Are Not the Compliance Test Method**

| Percentage of Time Out of Compliance | Percentage of Monitor Downtime | Appropriate Follow-Up Action |
| --- | --- | --- |
| < 2.0% | < 2.0% | If both cases exist, send letter acknowledging receipt of EER and encouraging proper O&M of CEM and facility. |
| > 2.0 and < 5.0% | > 2.0 and < 5.0% | If either or both cases exist, then warn by letter or telephone of unacceptable condition. |
| > 5.0 and < 10.0% | > 5.0 and < 10.0% | If either or both cases exist, then warn by letter of unacceptable condition, request explanations of condition, and request corrective action plan to prevent condition from reoccurring. |
| > 5.0 and < 10.0% for two consecutive quarters or > 10% | > 5.0 and < 10.0% for two consecutive quarters or > 10% | If either or both cases exist, then issue NOV and require performance (compliance) test for monitored pollutant, monitor certification (PS) tests, and request corrective action plan to prevent condition from reoccurring. |

*Reference:* Pfaff, R. O. 1989. Memorandum to state and local air program directors. EPA Region IV. Atlanta, GA.

## Level 2

If the EER report shows operational problems at the plant, a level 2 inspection may be specified in which the operations of the facility, its control equipment, and its CEM system undergo review.

A level 2 inspection may include both an examination of the emission control system operation and the emissions monitoring program. It is basically a systems audit and does not require the use of test equipment or gases.

## Level 3

Based on the findings and observations of the on-site inspection, it may be necessary to conduct limited testing on the CEM system to check that it is operating properly. Performance audit procedures are conducted, in which the CEM system is challenged with certified audit gases and/or filters to check data validity.

## Level 4

After a review and inspection of the plant CEM system and operations, the agency may require that a compliance test be performed. This next level of activity is used to determine source performance with respect to

authorization to emit 1 ton of $SO_2$ per year. Sources may trade (buy and sell) these allowances; however, the number of available allowances is controlled by the U.S. EPA. In order to build a new plant, a utility must accumulate the number of allowances equivalent to its expected yearly $SO_2$ emissions either by reducing $SO_2$ emissions from other plants or by buying or trading for the allowances. By limiting the number of allowances, it is expected that U.S. $SO_2$ emissions will be reduced by 10,000,000 tons per year from 1980 emission levels.

CEM systems play a central role in this program—by accounting for the number of allowances that can be bought, sold, or traded between the utilities. Because CEM systems are the basis and the standard for the scheme, it is important the CEM data be accurate; the costs associated with $SO_2$ emission allowances will enter into both operational planning and electrical dispatch programs. CEM data expressed in terms of mass emission rates (pounds of $SO_2$ per hour) are the basis for effecting these trades.

Although allowance trading is a new program, it does not eliminate previously established regulatory requirements. For example, if an NSPS emission rate is specified at 1.2 lb per $10^6$ Btu based on a three-hour rolling average, that rate must still be met, even though a source might accumulate allowances that would enable it to exceed this limit.

### Using CEM Systems to Provide Public Assurances

The public's increasing awareness of environmental pollution and the concern over the release of toxic pollutants into the air and groundwater has led to the evolution of a further application of CEM data. To help allay these concerns, requirements for continuous monitoring of stack emissions are increasingly being incorporated into regulatory permits. CEM systems are used to provide assurances to the public that a source is not emitting pollutants in excess of its standards *and* that the environmental control agency is watching that it does not.

Public-assurance CEM requirements often appear in state permits for hazardous waste incinerators and municipal waste combustors. In these permits, standards tend to be tighter than for other source categories and, frequently, more pollutants are required to be monitored. The data are generally used for direct compliance and may be telemetered to agency offices. The CEM system and the data that it will provide are, in effect, used as bargaining chips by both the source and the agency to gain some public acceptance for the siting and construction of new facilities.

TABLE 2-4    State CEM Program Options

| NSPS Delegation | SIP Rulemaking | Permits | Discretionary Programs | Implicit Programs |
|---|---|---|---|---|
| NSPS categories (Table 1) | Utilities Petroleum refineries $HNO_3$ $H_2SO_4$ Municipal waste combusters Other rules | PSD/NSR to construct and operate | Variances Orders Agreements | Guidelines QA Documents Control technology documents CEM manuals |

## STATE PROGRAMS

There are a number of methods that states can use to require CEM systems to be installed on a source. Also, as discussed previously, the CEM system data can be incorporated in various ways in air quality programs. Because of these options, agency regulatory policies, and work-force capabilities and limitations, CEM programs vary greatly between the states. Several states have very stringent programs based on CEM compliance monitoring for enforcement and public assurances. Other states, which have less stringent programs, accept U.S. EPA NSPS delegation for regulating new sources and have adopted minimum requirements for existing sources. Table 2-4 lists types of CEM program options available to the states.

### NSPS Delegation

Many states have received delegation by the U.S. EPA to regulate the categories of new sources for which new source performance standards have been established. In the case of NSPS sources for which CEM systems are required (Table 2-1) these states should have programs developed to assure that affected sources have installed the required CEM systems, to see that they are properly certified, and to receive and process excess emission reports and compliance data.

### State Implementation Plan Requirements (40 CFR 51)

Each state is required by the U.S. Clean Air Act to have a plan [the State Implementation Plan (SIP)] that will lead to the improvement and maintenance of air quality such that all areas of the state will meet national ambient air quality standards.

The U.S. EPA has established minimum CEM requirements for existing sources; these are given in Appendix P of Part 51 of the CFR. These minimum requirements address only four source categories: fossil-fuel-fired steam generators, sulfuric acid plants, nitric acid plants, and petroleum refineries. States were required to establish these regulations by 1976; however, not all states have done so (McCoy 1990). Even so, these particular minimum requirements only apply to the larger sources and provide for a number of exemptions, such as not requiring $SO_2$ monitoring unless $SO_2$ control systems are installed. More recently, existing municipal waste combusters (constructed before 1989) must also install CEM systems as a requirement of the SIP (U.S. EPA 1991).

State and local agencies, however, may impose more stringent requirements than the minimum requirements of Part 51 Appendix P. By applying established regulatory procedures, CEM systems can be required to be installed on other source categories or to be used directly for compliance. The development of such regulations proceeds through public hearings, combined with many approval processes. Under these procedures, it typically takes several years before a regulation is issued. As a result of such state initiatives and, now, the 1990 U.S. Clean Air Act Amendments, the requirements and exemptions of Appendix P have become somewhat outdated. New permit procedures, the allowance trading program, and new enforcement procedures resulting from the amendments will directly affect many existing sources.

### Operating Permits

States now issue operating permits for both existing and new stationary sources. In addition to being required to meet ambient air quality standards (NAAQS), states are required to prevent any deterioration of the air quality of already clean areas. The Prevention of Significant Deterioration (PSD) program addresses this requirement. Also, the operating permit program of Title V of the 1990 Clean Air Act Amendments will have a significant impact on emission sources.

An operating permit may include conditions that require the installation of CEM systems. These conditions may extend beyond those established for similar sources regulated under NSPS. Monitoring of gases such as HCl or volatile organic compounds may be required, or process cutoffs for excess emissions or real-time telemetry may be required in the permit. Sources often agree to very stringent CEM requirements to expedite approval of the operating permit.

## Discretionary Programs

Actions such as issuing permits give an agency a great deal of flexibility for achieving its goals. The ability of an agency to incorporate continuous monitoring requirements as part of these actions is within the agency's "discretionary authority." Discretionary programs include the following:

**Variances**    Under special circumstances a temporary authorization, called a variance, may allow a source to discharge pollutants in excess of the standard that would normally apply. A variance is generally issued for a specific time period, requiring that the source correct the problem and meet the standard by the end of the period. Special monitoring requirements are often included in the provisions of such variances.

**Orders**    Orders to comply with state agency requirements can be issued to a source under the enforcement authority given in Section 113 of the Clean Air Act. There are several types of orders: administrative orders, delayed compliance orders, and court orders. These differ with respect to the type of legal authority it takes to issue them.

**Agreements**    Agreements are often a result of bargaining between an agency and a source. Again, the purpose of any agreement is for the source to come into compliance with applicable regulations. As with orders, there are several types of agreements, each differing with respect to the level of authority or legal action exercised. Consent decrees, stipulation agreements, and court settlements are found in this group.

The use of discretionary authority allows fairness to be applied in regulatory programs. For example, a source may be able to gain time to install new control equipment by agreeing to the installation of a compliance CEM system. However, state discretionary authority can also result in arbitrary regulation, because it does not have the extensive public oversight that rule-making does.

## Implicit Programs

Either intentionally or unintentionally, CEM requirements are sometimes implemented through other than normal rule-making procedures. In order to provide guidance to its regulated sources or even to its own inspectors, a state may develop a "CEM Quality Assurance Manual" or "CEM Guidance Document." Such documents may contain rather specific information on how monitors should be installed, certified, calibrated, and audited. These documents may in turn be incorporated in source operating permits or agreements by "reference." That is, a permit may contain a clause stating that "all requirements of the '...Guideline Document' are incorporated herein [in the permit] by reference."

Such documents provide needed guidance, but they are not normally subject to agency rule-making processes. They often receive little external input and are not distributed for public comment. As a result, requirements that may be very stringent become established implicitly, not directly.

## TECHNICAL GUIDANCE BY THE U.S. EPA

The U.S. EPA provides policy guidance to state and local environmental control agencies through a number of avenues. There are four main areas where the technical resources of the U.S. EPA have been particularly valuable in developing CEM programs:

1. in setting standards (NSPS);
2. in developing and evaluating test methods and CEM certification procedures (performance specifications);
3. in establishing quality assurance criteria;
4. in providing enforcement guidance.

Guidance can be disseminated by the U.S. EPA headquarters directly to the states or through the EPA regional offices.

The degree to which guidance is provided to the states depends both upon state needs and the importance of the CEM programs for the regional offices. Program implementation that reaches the regulated sources comes most often through the state environmental control agencies, but may come through local agencies or, in special cases, through the EPA regional offices.

In addition to the establishment of CEM requirements through the new source performance standards, the federal EPA has also developed the set of CEM performance specifications used for CEM system certification and has mandated quality assurance programs for compliance CEM systems. Discussed only briefly here, these topics will be examined in detail in later chapters.

### Performance Specifications

Performance specifications give requirements for CEM system installation, instrument drift, and system accuracy, as well as other criteria. If a CEM system meets these specifications, it is considered capable of providing quality data for regulatory purposes.

The U.S. EPA does not approve specific brands of instrumentation or specific analytical methods (in the case of gas monitoring) for source monitoring systems. This is in contrast to the policy of approving specific instrument models for continuous *ambient* air analysis. The CEM performance specifications test procedures provide latitude in continuous monitoring system design and application by allowing sources to handle individual monitoring problems. Because stack conditions vary from plant to plant, a CEM system found to work well at one plant may not work at another. Vibration, heat, or other conditions may necessitate a careful analysis of available monitors and systems before one can be chosen for the job. The performance specifications define the procedures to be used for checking the *installed* instrument's performance. It is the installed instrument system, operating on the stack, that is approved, not the model.

Performance specifications (PS) have been established for seven classes of continuous emission monitors:

Performance specification 1—Opacity monitors
Performance specification 2—$SO_2$, $NO_x$ monitors
Performance specification 3—$O_2$, $CO_2$ monitors
Performance specification 4—CO monitors
Performance specification 5—TRS monitors
Performance specification 6—Rate (velocity) monitors
Performance specification 7—$H_2S$ monitors

These specifications are found in Appendix B of 40 CER60. States, however, may still require the installation of systems for which federal performance specifications have not been developed.

## Quality Assurance Programs

For the continuing operation of CEM systems, minimum quality assurance procedures are given by the U.S. EPA in Appendix F of 40 CFR 60. These include requirements for written plant CEM quality assurance plans and quarterly testing. The tests serve to provide an ongoing check of the original certification of the CEM system.

Appendix F of 40 CFR 60 applies only to those sources that use CEM systems for direct compliance. Sources under excess emission reporting programs technically are not required to meet the requirements of Appendix F. However, in order to obtain legally defensible data, it is essential for a source to develop its own quality assurance program or to adopt the directives of Appendix F. Many states have viewed CEM quality assurance as essential to their own programs and have patterned their quality assurance requirements after those of Appendix F.

Another U.S. EPA quality assurance program has developed traceability protocols for the direct comparison of CEM gases to standard reference materials (SRMs) of the National Institute of Standards and Technology (NIST) (U.S. EPA 1981). The availability and use of the "Protocol 1" gases prepared as a result of this program has done much to improve the quality of CEM data (Wright et al. 1987).

Both CEM system certification and quality assurance procedures will become more stringent for systems required to be installed under the 1990 Clean Air Act Amendments (U.S. EPA 1991). Rules promulgated for these systems are published in Part 75 of Title 40 of the *Code of Federal Regulations*.

### Enforcement Policy and Guidance

The U.S. EPA Stationary Source Compliance Division (SSCD) has actively promoted the use of CEM systems in its continuous compliance programs (Paley and Wright 1985). Through training programs and pilot projects, the division has provided guidance to the states in enforcement policy, the evaluation of performance specification tests, CEM auditing and QA program development, excess emission report evaluation, and inspection targeting. Continuous monitoring systems have become a key element in the U.S. EPA's enforcement programs.

## SUMMARY

Requirements for the installation of CEM systems at stationary sources arise from both federal and state regulatory programs. Whether state or federal, these programs specify pollutants that are to be monitored, emission limits, CEM system performance criteria, and reporting requirements.

Environmental control agencies have found the application of CEM systems to be valuable in the implementation of air quality programs. Whether used as operation and maintenance indicators for air pollution control equipment, as compliance monitors, or as the basis for allowance trades, CEM systems today are essential in achieving environmental goals.

### References

Kerstetter, D. L. 1985. Pennsylvania's continuous emission monitoring enforcement policy. In *Transactions—Continuous Emission Monitoring: Advances and Issues*. Air Pollution Control Association, Pittsburgh, pp. 187–198.

McCoy, P. G. 1985. The use of CEM data in Subpart D enforcement. In *Transactions—Continuous Emission Monitoring: Advances and Issues*. Air Pollution Control Association, Pittsburgh, pp. 146–174.

McCoy, P. G. 1990. The "CEM Nation" an analysis of U.S. EPA's database—1989. In *Proceedings—Continuous Emission Monitoring: Present and Future Applications*. Air and Waste Management Association, Pittsburgh, pp. 10–36.

McKee, H. C. 1974. Texas regulation requires control of opacity using instrumental measurements. *J. Air Pollut. Control Assoc.* 24:601–604.

Paley, L. R., and Wright, H. 1985. Use of continuous emission monitoring systems to implement EPA's continuous compliance program. In *Transactions—Continuous Emission Monitoring: Advances and Issues*. Air Pollution Control Association, Pittsburgh, pp. 128–145.

U.S. Environmental Protection Agency (U.S. EPA). 1981. *A Procedure for Establishing Traceability of Gas Mixtures to Certain National Bureau of Standards Standard Reference Materials*. EPA-600/7-81-010.

U.S. Environmental Protection Agency (U.S. EPA). 1986. Guidance: Enforcement applications of continuous emission monitoring system data. E. E. Reich to Regional Office Directors. In *EPA Air Programs Policy and Guidance Notebook*. OAQPS, Policy Notebook No. PN 113-86-04-22-030.

U.S. EPA. 1987. *Standards of Performance for New Stationary Sources*; *Quality Assurance Requirements for Gaseous Continuous Emission Monitoring Systems used for Compliance Determination*. 52 FR 21003 (June 4, 1987).

U.S. EPA. 1991. *Acid Rain Program: Permits, Allowance System, Continuous Emissions Monitoring, and Excess Emissions*. 56 FR 63002 (December 3, 1991).

U.S. Government. 1991. *Code of Federal Regulations*. Office of the Federal Register, Washington, DC.

Wright, R. S., Tew, E. L., Decker, C. E., von Lehmden, D. J., and Barnard, W. F. 1987. Performance audits of EPA protocol gases and inspection and maintenance gases. *J. Air Pollut. Control Assoc.* 37:284–285.

## Bibliography

Breton, H. 1990. Overview of German emission monitoring regulation. In *Proceedings—Specialty Conference on Continuous Emission Monitoring: Present and Future Applications*. Air and Waste Management Association, Pittsburgh, pp. 44–54.

DuBose, R. S. 1980. Monitoring continuous compliance of stationary air pollution sources. Paper presented at the TAPPI 1986 Environmental Conference, Denver; Transactions of the Pulp and Paper Industry, Proceedings, pp. 29–34.

Elman, B., Braine, B., and Stuebi, R. 1990. Acid rain emission allowances and future capacity growth in the electric utility industry. *J. Air & Waste Mgmt. Assoc.* 40:987–992.

General Accounting Office. 1990. *Improvements Needed in Detecting and Preventing Violations—U.S. GAO Report to the Chairman, Subcommittee on Oversight and Investigations—Committee on Energy and Commerce—House of Representatives*. GAO/RCED-90-155. U.S. General Accounting Office, Washington, DC.

Federal Minister for the Environment, Nature Conservation, and Nuclear Safety. 1988. *Air Pollution Control—Manual of Continuous Emission Monitoring*. Bundesministerium für Umwelt, Naturschutz und Reaktorsicherheit, Bonn, Germany (in English).

Helwig, V. G. 1980. Using continuous emission monitoring to document continuing compliance. Paper presented at the Air Pollution Control Association Meeting, Montreal. Paper 80-70.4.

Jahnke, J. A., and Aldina, G. J. 1979. *Continuous Air Pollution Source Monitoring Systems—Handbook*. EPA 625/6-798-005. U.S. Environmental Protection Agency, Research Triangle Park, NC.

Knauss, C. H. 1985. Continuous compliance—Which part of the elephant are you holding? In *Transactions—Continuous Emission Monitoring: Advances and Issues*. Air Pollution Control Association, Pittsburgh, pp. 222–225.

Lillis, E. J., and Schueneman, J. J. 1975. Continuous emission monitoring: Objectives and requirements. *J. Air Pollut. Control Assoc.* 25:804–809.

Makansi, J. 1991. Clean Air Act Amendments—The engineering response. *Power* 135(6):11–66.

McCoy, P. G. 1981. Continuous emission monitoring: A developing compliance tool. In *Proceedings—Specialty Conference on Continuous Emission Monitoring: Design, Operation and Experience*. Air Pollution Control Association, Pittsburgh, pp. 98–113.

Nader, J. S., Jaye, F., and Conner, W. 1974. *Performance Specifications for Stationary Source Monitoring Systems for Gases and Visible Emissions*. EPA 650/2-74-013.

Nazzaro, J. G. 1985. Continuous emission monitoring system approval, auditing and data processing in the Commonwealth of Pennsylvania. In *Transactions—Continuous Emission Monitoring: Advances and Issues*. Air Pollution Control Association, Pittsburgh, pp. 175–186.

Paley, L. R. 1980. Agency use of continuous emission monitoring data to ensure stationary source achievement of emission reductions. Paper presented at Air Pollution Control Association Meeting, Montreal. Paper 80-70.1.

Paley, L. R. 1981. Agency emphasis on continuous compliance programs—the impact upon emission monitoring methods. In *Proceedings—Specialty Conference on Continuous Emission Monitoring: Design, Operation and Experience*. Air Pollution Control Association, Pittsburgh, pp. 89–97.

Paley, L. R. 1982. Agency emphasis on continuous compliance programs: The impact upon emission monitoring methods. *J. Air Pollut. Control Assoc.* 32:705–708.

Paley, L. R. 1990. The effects of forthcoming acid deposition legislation on CEM systems. In *Proceedings—Specialty Conference on Continuous Emission Monitoring: Present and Future Applications*. Air and Waste Management Association, Pittsburgh, pp. 370–379.

Quarles, P., and Wayne, A. 1985. The Missouri continuous emissions monitoring pilot project: Preliminary findings and recommendations involving opacity excess emission reports and reporting practices in Missouri. In *Transactions—Continuous Emission Monitoring: Advances and Issues*. Air Pollution Control Association, Pittsburgh, pp. 146–158.

Reich, E. E. 1985. CEMS programs of the EPA stationary source compliance division. In *Transactions—Continuous Emission Monitoring: Advances and Issues*. Air Pollution Control Association, Pittsburgh, pp. 64–65.

U.S. Environmental Protection Agency (U.S. EPA). 1971. Standards of performance for new stationary sources. 36 FR 24876 (December 23, 1971).

U.S. EPA. 1975. Requirements for submittal of implementation plans—standards for new stationary sources—emission monitoring. 40 FR 46240 (October 6, 1975).

U.S. EPA. 1977. Procedure for NBS-traceable certification of compressed gas working standards used for calibration and audit of continuous source emission monitors (revised traceability protocol no. 1). In *Quality Assurance Handbook for Air Pollution Measurement Systems*, Vol. 3, *Stationary Source Specific Methods*. EPA-600/4-77-027b. Section 3.0.4.

U.S. EPA. 1979. Standards of performance for new stationary sources; continuous monitoring performance specifications. 44 FR 58602 (October 10, 1979).

U.S. EPA. 1985. *Air Compliance Inspection Manual*. EPA-340/1-85-020.

U.S. EPA. 1991b. Standards of performance for new stationary sources—Appendix B—performance specifications. *U.S. Code of Federal Regulations*. U.S. Government Printing Office, Washington, DC.

U.S. EPA—Office of Air Quality Planning and Standards (OAQPS). *EPA Air Programs Policy and Guidance Notebook*. Research Triangle Park, NC. This is an ongoing notebook in which the following dated documents, among others, have been included:

**1982.** Memorandum: Guidance concerning EPA's use of continuous emission monitoring data: K. M. Bennett to Regional Office Directors. In *EPA Air Programs Policy and Guidance Notebook*. Policy Notebook No. PN 113-82-08-12-014.

**1984.** Final technical guidance on the review and use of excess emission reports: E. E. Reich to Regional Office Directors. In *EPA Air Programs Policy and Guidance Notebook*. Policy Notebook No. PN 113-84-10-05-021.

**1984.** Guidance on federally-reportable violations for stationary air sources: J. C. Potter to Regional Office Directors. In *EPA Air Programs Policy and Guidance Notebook*. Policy Notebook No. PN 113-86-04-11-029.

**1988.** Compliance monitoring strategy for FY 1989: J. S. Seitz to Regional Office Directors. In *EPA Air Programs Policy and Guidance Notebook*. Policy Notebook No. PN 114-88-03-31-006.

**1988.** Transmittal of reissued OAQPS CEMS policy: G. A. Emison to Regional Office Directors. In *EPA Air Programs Policy and Guidance Notebook*. Policy Notebook No. PN 113-88-03-31-048.

**1988.** Transmittal of $SO_2$ continuous compliance strategy: J. S. Seitz to Regional Office Directors. In *EPA Air Programs Policy and Guidance Notebook*. Policy Notebook No. PN 113-88-07-05-051.

**1988.** Transmittal of OAQPS interim control policy statement: J. S. Seitz to Regional Office Directors. In *EPA Air Programs Policy and Guidance Notebook*. Policy Notebook No. PN 113-88-03-31-047.

von Lehmden, E. J. 1981. Quality assurance requirements for gas continuous monitoring systems for compliance. In *Proceedings—Specialty Conference on Continuous Emission Monitoring: Design, Operation and Experience*. Air Pollution Control Association, Pittsburgh, p. 88.

Walton, J. W., Stewart, J. W., and Singleton, F. O. 1985. Utilization of continuous emission monitoring data for enforcement purposes. In *Transactions—Continuous Emission Monitoring: Advances and Issues*. Air Pollution Control Association, Pittsburgh, pp. 19–207.

Willenberg, J. M. 1990. Washington State's CEMS experience and enforcement guidance. In *Proceedings—Specialty Conference on Continuous Emission Monitoring: Present and Future Applications*. Air and Waste Management Association, Pittsburgh, pp. 37–43.

Wright, R. S., Wall, C. V., Decker, C. E., and von Lehmden, D. J. 1989. Accuracy assessment of EPA protocol gases in 1988. *J. Air & Waste Mgmt. Assoc.* 39(9):1225–1227.

# 3

# Extractive System Design

Analyzers used to monitor stack gases must be incorporated into a system — one that can protect the analyzer from the plant environment and provide a representative stack-gas sample for analysis. Either in-situ or extractive systems can be designed to do this. Extractive systems were originally developed for continuous emission monitoring and will be discussed in this chapter; in-situ systems will be discussed later.

Delivering and conditioning a hot, moisture-saturated, particle-laden flue gas for analysis can be a difficult process. Because the pollutant gases may be lost in this process, it is critical that the extractive system be properly designed and manufactured. Extractive system designs are associated with specific CEM system vendors, having been developed through years of industrial applications experience. Although it is possible and relatively easy to design and construct an extractive system from basic components, most systems purchased today are provided by vendors on a turn-key basis, that is, the system is provided from start to finish. The two basic types of extractive systems that vendors offer are (1) fully extractive, source-level systems and (2) dilution systems. Each system has its own advantages and disadvantages in any given application. There is no "best system" because plant and flue-gas conditions vary widely.

This chapter describes both source-level and dilution extractive systems and the components that are used in their design. Knowledge of these extractive techniques will assist in evaluating proposed system designs or systems that have already been installed.

## SOURCE-LEVEL EXTRACTIVE SYSTEMS

Source-level extractive systems remove gas directly from the stack or duct, filter out particulate matter, and transport the gas for analysis. Three types of source-level extractive systems are marketed commercially:

1. hot–wet systems (Figure 3-1)
2. cool–dry systems with conditioning at the probe (Figure 3-2)
3. cool–dry systems with conditioning at the CEM system shelter (Figure 3-3)

### Hot–Wet Systems

The simplest type of extractive system uses a heated line to transport the flue gas to an analyzer that incorporates a sample cell heated above the flue-gas temperature. The gas is delivered to the analyzer both hot and wet, but minimally conditions the gas by removing particulate matter with a coarse filter located at the probe. The technique is useful when monitor-

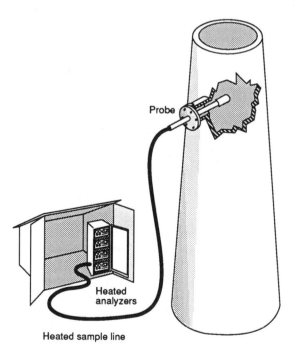

**FIGURE 3-1.**    A hot and wet CEM system.

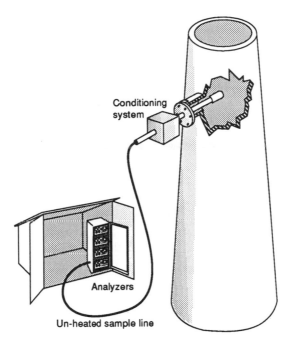

**FIGURE 3-2.**    A cool and dry CEM system with conditioning after the probe.

ing of water-soluble gases, such as HCl and $NH_3$, is required or when emission values are to be reported on a wet basis. Because water is not removed from the sample gas, problems associated with condensation systems are avoided. However, care must be exercised in maintaining the temperature of the sample above the dew point, from the probe to the analyzer exhaust. If the heating system should fail, moisture will readily condense and foul the system. This can lead to corrosion, plugging, or damage to the analyzer.

The hot–wet systems were marketed successfully in the 1970s in association with ultraviolet analyzers designed for the measurement of $SO_2$ and NO. The technique is used today by several systems vendors for the applications noted previously.

## Cool–Dry Systems

In a more widely used extractive system design, the gas is conditioned before it enters the analyzer. The gas temperature is reduced to ambient temperature and moisture is removed, so that the sample is both cool and

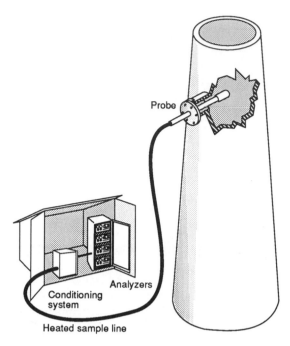

Probe

Analyzers

Conditioning
system

Heated sample line

**FIGURE 3-3.**    A cool and dry CEM system with conditioning at the CEM system shelter.

dry. The conditioning may be conducted either at the probe location (Figure 3-2) or at the analyzer shelter (Figure 3-3). Conditioning at the probe location offers the advantage of using unheated sampling line; however, daily preventive maintenance of the conditioning system at the probe may be inconvenient. Conditioning at the shelter or CEM room enables the CEM system operator to check system performance conveniently. However, the necessary heat-traced lines must be maintained at a proper temperature over their entire length.

Extractive systems that condition the flue gas allow greater flexibility in the choice of analyzers and are commonly used when emission calculations are performed on a dry basis or when a monitoring of a number of different gases is required. Although this type of system is not as sophisticated as some others, it is flexible enough to accommodate engineering changes when applications problems arise. In problem applications, the system components can be readily modified or replaced so that the system can meet performance specifications.

A source-level extractive system is made up of a set of basic components: probe, sample line, filters, moisture removal system, and pump.

Because the operation of an extractive system is dependent upon the design and quality of each component, as well as on their arrangement in the system, it is necessary to review these characteristics.

## Sample Probes

A simple probe can be made by merely inserting an open metal tube into the stack or duct. This may be adequate in sampling situations where no particulate matter is present. However, flue gases free of particulate matter do not often occur in those sources that are subject to CEM regulation. An open tube can be easily plugged when particulate matter is present, especially if the flue gas has a high moisture content. Also, water may condense and combine with the particulate matter to produce an agglomerated material that can plug the probe more readily. To minimize such problems, a filter can be placed on the end of the probe (Figures 3-4a–c).

Filters made of sintered stainless steel and porous ceramic materials are commonly used to prevent particles from entering the sample tube. Sintered metal is made by compressing micrometer-sized metal granules

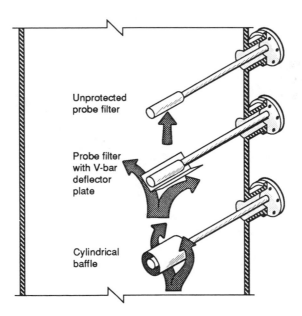

**FIGURE 3-4.**    (a) A simple probe filter. (b) Sintered filter with a baffle plate deflector. (c) Sintered filter with a cylindrical deflector.

under high pressure and elevated temperatures. The metal fuses and acquires a porosity that is dependent upon the compression pressure. Sintered stainless steel filters that are capable of filtering out particles from 10 to 50 $\mu$m in size are frequently used as probe filters. Some systems use filters that exclude particles greater than 1–2 $\mu$m in size, but the finer the filter, the more difficult it will be to draw sample gas through the filter, and pump capacity will need to be increased.

Sintered filters or ceramic filters can become plugged by particles impacting on and penetrating into the porous material. To minimize plugging, a baffle plate can be attached to the filter to deflect particles from the filter surface (Figure 3-4b). Particles will then follow streamlines generated by the plate, whereas the pollutant gases will still diffuse into the probe. Another way to minimize plugging is to attach a cylindrical sheath around the filter (Figure 3-4c). Gas will diffuse into the annular space between the filter and baffle, but most particles will not be able to make the 90° change in direction to enter the space. If the end of such a sheath is partially closed off with a plate having holes drilled into it to provide for gas diffusion, the external filter probe can undergo a probe calibration check. Calibration gas can be injected into the sheath to flood out the stack gas during the calibration intervals.

In the probe designs shown in Figure 3-4a–c, the coarse particulate filter is external to the probe. In other probes, this filter is mounted internally, in a housing mounted outside the stack (Figure 3-5).

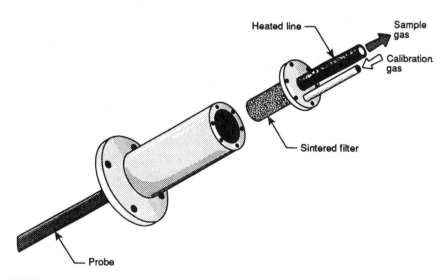

**FIGURE 3-5.**    An internal coarse filter.

In such an internal filter assembly, a heater can be either fitted into the assembly or placed around the outside of the filter holder. This enables hot gas to be drawn through the filter and passed through to the heated sample line and the remainder of the conditioning system. The principal advantage of this configuration is that the coarse filter can be easily unclamped and inspected. The entire probe assembly does not need to be unbolted and removed from the stack to replace the filter, and if the probe should become plugged with particulate matter, the plug can be pushed out with a rod. Also, if the probe is mounted at an angle (Figure 3-5), water or acid condensed in the probe can roll back into the stack.

Extractive systems designed with internal filters can be readily calibrated at the probe filter housing. Calibration gases can be injected at the top of the housing to flood the annular space between the filter and the housing and give a check of system integrity from the filter back to the analyzer. This is difficult to do with external filter systems because the filter is more subject to the flue gas flow and excessive amounts of calibration gas may be needed to flood the area around the filter.

There are other variations of the designs shown in Figure 3-5 that are used. One variation utilizes a coarse filter (10–50-$\mu$m porosity) at the probe tip, but incorporates an internal fine filter (1-$\mu$m pore size) at the flange assembly. Another variant uses a bellows valve to close off the internal filter from the probe to reduce the amount of gas necessary to perform a probe calibration.

A major problem associated with the probes illustrated in Figures 3-4a–c and 3-5 is that the sintered filters still can plug with particulate matter. Another system, designed to minimize this problem, utilizes an inertial filter. The inertial filter is an internal filter design that can be incorporated either in the probe or in a system external to the probe (Figure 3-6).

In the inertial filter system, a pump (usually an eductor pump) pulls the flue gas through a cylindrical filter. The internal filter can be made of sintered stainless steel or porous ceramic, usually having a pore size of 2–5 $\mu$m. As the flue gas moves down the tube, a sample is pulled off from the filter at a direction perpendicular to the gas flow (radially) using another sampling pump. The gas velocity through the tube can vary from 70 to 100 ft/s, whereas the radial velocity of the gas pulled from the filter may be only 0.005 ft/s. The large particles in the flue gas are swept through the tube because of their inertia in the gas stream and are exhausted back into the stack. Because of the low radial sample velocity, larger particles are also less likely to break from their streamlines and enter the filter. The 70–100-ft/s flow also aids in sweeping particles off of the filter surface back into the gas stream.

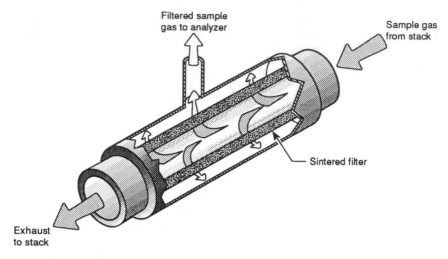

**FIGURE 3-6.**    The inertial filter.

Conceptually, this system may appear ideal but, actually, submicron-sized particles (< 1-$\mu$m diameter) will follow the radial sample flow and enter the tubular filter. These embedded particles will further assist in the filtering action, reducing the filter pore size and removing particles down to 0.5-$\mu$m diameter from the sample stream. However, this also means that the filter can eventually become plugged.

Filter plugging can be a problem with any fully extractive system. Plugging can be minimized by "blowing back" on the filter using high-pressure gas, plant air, or steam—air at pressures from 60 to 100 pounds per square inch (psi) is blown back through the filter, opposite to the normal direction of gas flow. The blowback can be pulled by first pressurizing a surge tank and suddenly releasing the pressure to shock the particulate matter out of the pores of the filter. Depending upon the particle characteristics and concentration, filters are blown back at intervals of 15 min to 8 h for durations of 5–10 s. Fifteen-minute blowback cycles are common. Care must be taken in the blowback system so that the blowback gas does not cool the probe to the extent that acids or other gases condense.

**Sample Line**

A sample line is used to transport the sample from the probe to other elements in the conditioning system or to the analyzer. In some cases [where water-soluble or condensable gases such as HCl and volatile organic compounds (VOCs) are to be analyzed], the gases are delivered hot, at flue-gas temperature or higher, using a hot–wet system. It is more common, however, to use a heated sample line to send the gas to a moisture removal system that cools the gas and removes water contained in the sample gas stream.

Moisture removal can be performed either at the probe or at the analyzer location. If the removal is done at the probe, unheated sample line (or sample line with a lower temperature rating—freeze protection line) can be used to transport the sample to the analyzer. If moisture is removed at the analyzer location, heated sample line must be used all the way from the probe to the moisture removal system. This distance can be as little as 2–3 ft for stack-located systems, but is more commonly 100–250 ft. It is also important to insulate and/or heat the junctions between the heated sample line and the probe and the sample line and the moisture removal system, otherwise cold spots can develop that may eventually cause plugging and corrosion.

The time it takes for gas to flow through a sample line depends on the sample tube diameter, its length, and the flow rate. This lag can be roughly calculated from the expression

$$t = \frac{V}{Q_{sl}} \qquad (3\text{-}1)$$

where $t$ = the lag time (in minutes)

$V$ = sample line volume (in liters)

$Q_{sl}$ = volumetric flow rate of gas through the line (liters per minute)

(McNulty et al. 1974). This expression is oversimplified because it assumes that there are no frictional wall effects restricting the flow. It shows, however, that if the sample line length is increased, the lag time will increase as the sample line volume increases. Typically, for a flow rate of 1 standard liter per minute (1/min), the lag time for a 100-ft section of 0.25-in. internal diameter tube at a pressure drop of 152 mm Hg is only 30 s.

There are other problems associated with long sample lines that have a more practical nature. Uniform temperatures are difficult to maintain over long sample lines, and condensation of water or acid gases in cooler "pockets" can lead to increased system corrosion or line plugging if fine particles are not adequately filtered at the probe. Plugs of particulate matter are difficult to remove in heated sample line and if a heater wire should break or turn out, it is often difficult to find the break. Sample lines can also become contaminated with condensable or reacted materials, which become difficult to remove. On occasion, entire sample lines will need to be replaced because these problems cannot be resolved. It is evident, then, that a good source-level extractive system design will minimize the length of heated sample line.

In properly installed systems, heated sample lines are generally less than 250 ft in length. They slope at a minimum of 5° from the probe all the way to the moisture removal system or analyzer. Because sampling line can distort when heated, care is taken to avoid sags in the line, where condensate might collect. Hot spots are also avoided by preventing the line from touching itself or other sample lines.

Sample line is generally incorporated in a tube bundle, or umbilical. A typical umbilical tube assembly is shown in Figure 3-7. Sample lines can be operated either at a constant power density or can be self-limiting (self-regulating). The self-limiting lines are more appropriate to CEM applications because the line will maintain a specified minimum operating temperature if the ambient temperature should fall (e.g., during subzero conditions in winter) and will not exceed a specified maximum temperature if the ambient temperature rises. The sample line, blowback line, and zero-gas–calibration lines as well as electrical cables can all be incorporated into the umbilical. Also, thermocouples can be placed at intervals

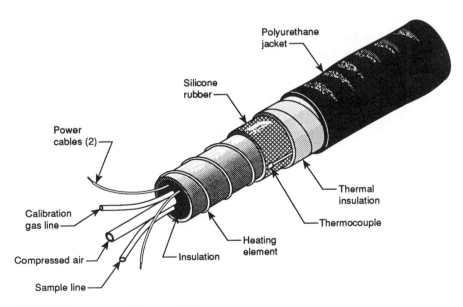

**FIGURE 3-7.**    An umbilical assembly.

along the line to monitor the temperature and warn of any temperature reduction. Sample lines are generally custom-made and can contribute significantly to the expense of an extractive system. Sample line is usually priced per running foot and a total fixed price generally is not quoted by CEM system vendors unless the exact length is provided in a CEM system bid specification.

Although much information has been catalogued on the relative chemical resistance of sample line materials to corrosive gases (McNulty et al. 1974, Podlenski et al. 1984) most systems today use either PFA Teflon or No. 316 stainless steel to avoid chemical attack and wall adsorption effects. Problems do occur with Teflon, because its softening temperature is near 250°C. Efforts used to increase sample line temperature to avoid condensation sometimes result in melting the line and umbilical cable.

### Moisture Removal Methods

Moisture is usually removed before the hot flue gas enters a sample pump because water vapor and acid gases can easily condense in an unheated pump and corrode the interior. Cooling beyond the dew point (the temperature at which air is saturated with moisture) will cause moisture to condense, and many of the moisture removal systems are designed to reduce the sample temperature below this point. Condensation systems and permeation dryers are most commonly used for this purpose in extractive systems. A successful system is correctly sized for the expected sample-gas flow rates and stack-gas moisture content; also, the system must remove condensed water rapidly from the gas stream to minimize contact with dried gas.

### *Condensation Systems*

A typical approach to moisture removal is to use a mechanically refrigerated condenser such as that shown in Figure 3-8. This condensation system consists of a coil of glass of Teflon placed in a liquid (the liquid can be water, an antifreeze solution, or in some cases air) cooled by a refrigeration system. To avoid freezing water in the coil, the temperature is not allowed to go below 35°F. A liquid trap is used to collect the condensed water vapor, and the liquid is either periodically or continuously removed using a peristaltic pump. Condensed water usually is removed automatically because, in a manual operation, if the CEM system operator fails to perform this maintenance task, there exists a significant risk that the trap will fill and water will flood into the sample line.

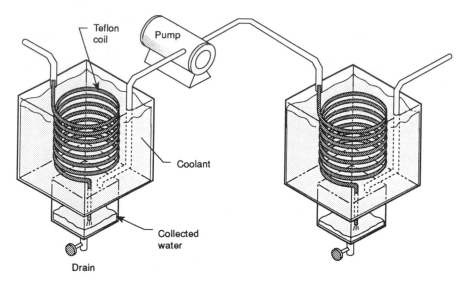

**FIGURE 3-8.**    Refrigerated condenser moisture removal system with a secondary chiller under positive pressure.

Another coil can be added to reduce the water content further, but a more efficient technique is to place the sample pump after the first coil and to transport the gas from the chiller, under pressure, to a second chiller as shown in Figure 3-8. A gas under pressure will condense more readily than a gas under vacuum. Because it is more difficult, when the gas is under pressure, for the molecules to escape the liquid surface vaporize, this pressurization will, in effect, reduce the moisture content to a level lower than it would be under atmospheric pressure.

The traditional method of chilling a condenser coil or tube has been to use a refrigerator, with a compressor pump and a circulating liquid to increase heat transfer. In refrigerator systems, Freon sometimes escapes, pumps wear out, or algae start growing in the liquid. Vortex coolers avoid these problems by using a counterflow of high-pressure plant air to cool the sample gas. However, the power requirements necessary to generate the high flow rates of air needed may be excessive in many applications.

Another approach is found in thermoelectric coolers that take advantage of the Peltier effect to chill the sample gas. The Peltier effect occurs when two dissimilar metals or semiconductors are joined in a loop and a voltage source generates a current through the loop. Because of the differing electron distributions in the dissimilar materials, a small voltage difference exists across their junction. The junction will heat up where the voltage difference opposes the voltage difference of the battery. At the other junction, thermal energy is absorbed from the surroundings and is converted to electrical energy in order to balance the electron flow. This absorption, of course, reduces the temperature of the surroundings.

Semiconductor thermoelectric coolers are made as flat plate heat exchangers in portable, compact arrangements and are seeing increased use in sample-conditioning systems. These solid-state devices have few moving parts except perhaps for a fan and peristaltic pump. They can also be designed for rapid separation of the dry gas from the condensed water vapor.

A frequent comment made about condensation-type moisture removal systems is that pollutant gases are absorbed in the condensate. This can occur easily for gases such as HCl and $NH_3$, which have high solubility in water. However, gases such as $SO_2$ and NO are less soluble. Also, this solubility is reduced as the acidity of the condensate increases. At source-level concentrations on the order of 100 ppm or greater, solubility of $SO_2$ and NO usually has not been a problem with respect to systems meeting U.S. EPA relative accuracy requirements. However, with the advent of more-stringent emission standards, resulting in requirements for the measurement of gases in the range of 10 ppm, more care must be taken in system design to avoid absorption by water. Unfortunately there is little published data on the effects of condensate absorption for commercial systems. A theoretical framework for such determinations is provided by McNulty et al. (1974) for those who wish to pursue this issue.

Systems (such as jet stream condensers) have been developed for more rapid chilling and separation of moisture from the sample gas. These condensers rapidly change the direction of flow of the gas through a cooled impinger (Figure 3-9).

### Permeation Dryers

Permeation dryers take advantage of a unique property of ion exchange membranes, a class of synthetic plastic materials that allows for the transport of specific molecular species. One such material, Nafion, will transmit water vapor if the partial pressure of the water vapor is different on each side of the membrane. A typical permeation dryer (Kertzman 1973) uses a bundle of Nafion tubes as shown in Figure 3-10. In this permeation dryer assembly, the wet sample gas enters the tubes and dry purge gas flows in the opposite direction on the outside of the tubes. The driving force in this exchange also can be provided by either evacuating the shell side of the dryer or by back-purging with dry gas taken from the tube exit. The permeation dryer requires that the gas be held above the dew point upon entering the dryer. The drying efficiency increases with

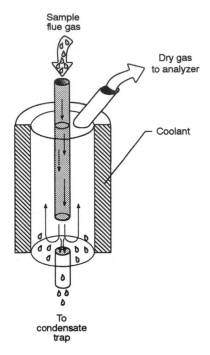

**FIGURE 3-9.**    A jet stream condenser.

the length of the tubes and is also dependent upon the sample inlet and upurge gas pressures. Capacity can be increased by increasing the number of tubes in the bundle.

The permeation dryer offers a number of advantages over refrigerated chillers because no mechanical parts are incorporated in the system, no condensate trap is required, and the question of pollution absorption in condensed water is avoided. However, the system is prone to plugging, either from droplets of condensed material or from particulate matter introduced by improperly filtered samples. Problems of condensing liquids may be minimized by heating the entrance side of the dryer.

### Miscellaneous Drying Techniques
Other drying techniques have been used or attempted in extractive monitoring systems. Cyclone-type devices installed in or near the probe, coalescing filters, "knock-out jars," and other engineering afterthoughts are frequently encountered.

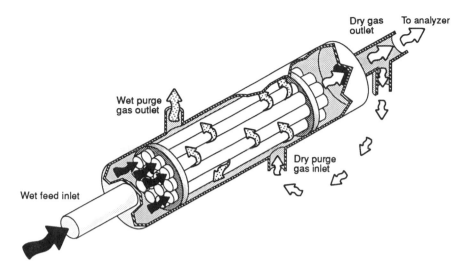

**FIGURE 3-10.**    A permeation dryer assembly.

The use of chemical desiccants [such as calcium chloride $(CaCl_2)$, concentrated sulfuric acid $(H_2SO_4)$, calcium sulfate $(CaSo_4)$, etc.] to remove moisture is not common in CEM systems. Because desiccants have to be periodically regenerated or replaced, they are considered to present an unnecessary maintenance task. Also, to justify their use, it must be shown that the gases being measured do not react with, or adsorb or absorb with the material.

### Sampling Pumps

The sample pump is an important element of the extractive system and is used to transport the sample from the stack to the analyzer. A pump should be sized appropriately to meet the demands of the gas analyzers and be designed so that no air inleakage occurs (i.e., around a rotary shaft seal) and no contamination is introduced from pump lubricating oils. Two types of pump that meet these criteria are (1) the diaphragm pump and (2) the ejector pump. These pumps are commonly used in source monitoring applications.

The diaphragm pump operates by mechanically stroking a piston or connecting rod to move a flexible diaphragm (Figure 3-11). The diaphragm is circular and can be made of flexible metal plate, Teflon, polyurethane, or other type of elastomer. The reciprocating action of the diaphragm moves the gas in short bursts. As the diaphragm is raised, gas is drawn

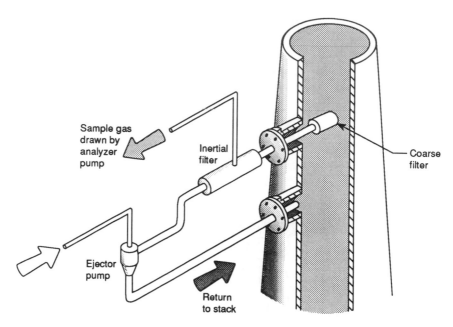

**FIGURE 3-13.**    An externally mounted ejector pump.

## Fine Filters

The coarse filter is used to remove larger particles from the sample gas. Because the majority of gas analyzers require almost complete removal of particles larger than 0.5 $\mu$m, additional filtration is necessary. This is accompanied by incorporating a fine filter before the analyzer inlet. The location of the fine filter is an important consideration in the CEM system design and is dependent upon the susceptibility of the other system components to the effects of fine particles. There are two groups of fine filters: surface filters and depth filters.

A surface filter can be simply a filter paper that excludes particles of a certain size. The filter material is porous to the moving gas, but the pores are of such a size that they prevent penetration of the fine particulate matter. A filter cake can also build up on the filter, further reducing the size of particles passing into the gas stream. Because of the filter cake and developed electrostatic charges, surface filters can remove particles smaller than the actual filter pore size.

Depth filters collect particles within the bulk of a filter material. The filter may consist of loosely packed fibers of quartz wool or of filter material wrapped to a depth sufficient to remove the fine particles. Such filters work particularly well for dry solids and moist gas streams containing aerosols. A depth filter can also be used as a probe filter, a technique that is used in several German extractive systems.

## DILUTION SYSTEMS

The main problem associated with source-level extractive systems is the need to filter and condition relatively large volumes of stack gas. This problem can be largely avoided by using dilution systems where gas is drawn into the probe at low flow rates, sometimes two orders of magnitude less than in a source-level system (e.g., 0.1 l/min vs. 10 l/min). This means that there will be less moisture to remove and less particulate matter to filter. Because the flow rate is relatively low, particles are more likely to follow the flue-gas streamlines than to enter the probe.

Dilution systems are used in conjunction with ambient air analyzers; a feature that can provide significant advantages to a source that has had previous experience with ambient air analyzers or that is operating an ambient air network. In case of analyzer problems, the CEM system

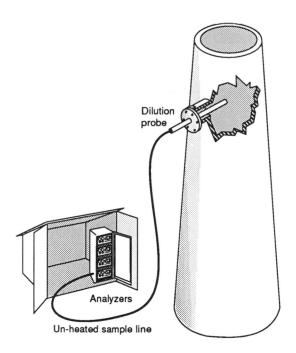

**FIGURE 3-14.**    A dilution probe CEM system.

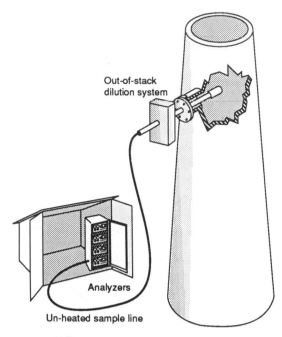

Out-of-stack
dilution system

Analyzers

Un-heated sample line

**FIGURE 3-15.**    An external dilution CEM system.

analyzers could be swapped for ambient analyzers or spares maintained
for both purposes. Plant technicians may already be familiar with the
operation and maintenance needs of the analyzers and may not require
additional training.

There are two approaches to designing dilution systems. One is to
dilute the stack gas inside the sample probe and bring the sample down to
the analyzer using unheated line (Figure 3-14). Another is to dilute the
stack gas outside the stack (Figure 3-15). As with fully extractive system
conditioning, the out-of-stack dilution can be performed either at the
stack or at the CEM shelter. If diluted at the stack, unheated sample line
can be used to transport the gas to the analyzers, but if diluted at the
shelter, heat-traced line must be used.

### Dilution Probes

A dilution probe dilutes the stack gas in the probe to such a degree that
the dew point of the diluted gas will be less than the lowest ambient
temperature at the sampling location. This enables the CEM system to
avoid the use of heat-traced line and simplifies the gas transport system.

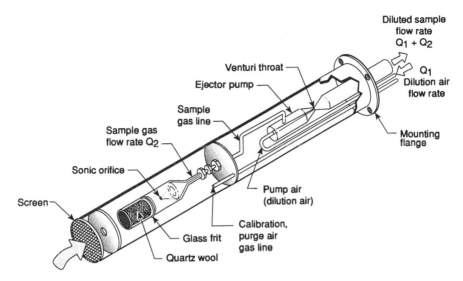

**FIGURE 3-16.**    The dilution probe.

One of the first and most successful dilution probes utilizes a sonic orifice coupled with an ejector pump designed into the probe body (Figure 3-16). This probe was originally developed in the Netherlands (Bergshoeff and van Ijssel 1978) and has seen many successful applications (e.g., Frank and Mullowney 1990; Maurice, Robertson, and Howder 1986). The ejector pump, operates at flow rates of 1–10 l/min. A glass sonic orifice (consisting of a glass tube drawn to a point, as shown in Figure 3-16) is chosen to limit the flow of sample gas to flow rates from 50 to 500 ml/min. The condition for obtaining a critical flow for the glass sonic orifice is that the ratio of the absolute pressure at the venturi throat to the stack static pressure must be less than or equal to 0.53 (Brouwers and Verdoorn 1990). The dilution ratio $R$ is determined by calculating the following:

$$R = \frac{Q_1 + Q_2}{Q_2}$$ (3-2)

where $Q_1$ = dilution air flow rate (in liters per-minute)
    $Q_2$ = sample gas flow rate (in liters per minute)

Dilution ratios of 100 to 1 are typical, although higher ratios are used for hot, saturated gas streams. In coupling the dilution probe to an ambient air analyzer, attention must be paid to the range of the analyzer. If the lowest instrument range should be 0–5 ppm and it is required to measure a pollutant stack gas at a nominal concentration of 50 ppm of the pollutant, a 100-to-1 dilution ratio would provide a sample to the analyzer of 0.5 ppm. This would be at the low end of the range, where the analyzer sensitivity is lowest. If the instrument noise or drift is high at this part of the scale, it could be very difficult for the system to pass a relative accuracy test.

Although the dilution probe has been successfully applied, it is not the solution to all extractive sampling problems. In cases where wet, caking, or sticky particulate matter is present, the probe can still become plugged even though it is pulling at a low flow rate. Also, water droplets in the flue gas stream may cause problems, especially if the droplets wet the glass wool filter or enter the orifice. Under normal conditions, when the probe is heated, the glass capillary will not become plugged, but if it does, it must be replaced or cleaned.

The dilution probe is sensitive to changes in stack pressures and temperatures (Myers 1986). In installations where the stack static pressure is highly negative ($< -10$ in. $H_2O$), the venturi vacuum may not be sufficient to overcome the pull of the stack negative pressure. Also, for a change of stack pressure of 4.1 in. $H_2O$ (from the pressure at which the probe was calibrated), the reading will change by 1%. This may contribute a significant error in some facilities, but the data can be corrected easily if the pressure is monitored also.

It should also be noted that air is used for the dilution. This prevents the use of an oxygen analyzer in the CEM system, because the contribution of stack gas oxygen to the sample would be swamped by the background 21% oxygen levels of the dilution air. Most commercial oxygen analyzers are not designed to measure differences in oxygen levels at diluted values of 0.1–1%. If it is required to correct pollutant emission data for stack dilution air, a $CO_2$ analyzer is used in the system. The air must be clean and free of any of the gases being measured, or significant errors can occur. The quality of vendor air clean-up systems should be carefully evaluated when considering the purchase of a dilution system. A CEM system that analyzes a flue gas diluted with dry air is virtually measuring a dry gas. There is proportionately little water in the diluted sample and the water removal requirement for the conditioning system, in some cases, may be dispensed with.

Another type of dilution probe uses a "loop dilution" technique (Whitt, Defriez, and Collier 1985). Developed by H. Defriez, this unique system incrementally dilutes the sample gas to an equilibrium value by removing water vapor with a Peltier cooler at the stack exit and diluting the incoming stack gas with the dried gas.

## Out-of-Stack Dilution Systems

An alternative approach to dilution is to dilute the gas at an assembly mounted after the probe, on the stack flange, or at the CEM shelter (Figure 3-16). In this configuration, the gas can be diluted using either a metering pump or critical orifice. One of the advantages of this approach is that undiluted sample can be sent first to an oxygen analyzer before it proceeds to the dilution system. This provides an alternative to using a $CO_2$ analyzer for the diluent monitor, which is necessary in a dilution probe system. If the gas is diluted in the CEM shelter, a heated sample line must be used to transport the gas. Then, one of the principle advantages of using a dilution system is lost.

## Dilution Systems—Wet or Dry Measurement?

A question that frequently arises when considering the use of a dilution system is whether the emission values are given on a wet basis or a dry basis. Dilution systems that do not dry the diluted sample gas give concentration valves (ppm) on a wet basis. If there is no moisture removal, there will still be moisture in the diluted sample. When the diluted concentration (measured with ambient air analyzer) is scaled back up by multiplying it by the dilution ratio, the moisture content is scaled back up also. The concentration of the pollutant gas in the original wet sample is obtained.

Consider an example of 1 ml of flue gas that contains 0.1 ml of water vapor (10% $H_2O$) and is diluted with 99 ml of dry air (Figure 3-17). In the diluted sample, the percentage of water vapor is reduced to about 0.1% of the total volume. Essentially, the analyzer is measuring $SO_2$ in a background gas containing 99.9% dry gas and 0.1% water vapor. Therefore, the flue-gas $SO_2$ concentration obtained is given on a wet basis. When the diluted concentration is scaled back up by multiplying it by the dilution ratio, the concentration of $SO_2$ in the original wet sample is obtained.

Consider again, from Figure 3-17, that the diluted $SO_2$ concentration is measured to be 3 ppm. This means that for every 100 ml of gas analyzed, the volume of $SO_2$ equals $3 \times 10^{-6} \times 100 = 3 \times 10^{-4}$ ml. However, because no $SO_2$ is in the dry air used for dilution, this must be the volume of $SO_2$ gas contained in the 1 ml of flue gas drawn into the dilution system. A volume of $3 \times 10^{-4}$ ml in 1 ml is equivalent to a flue-gas

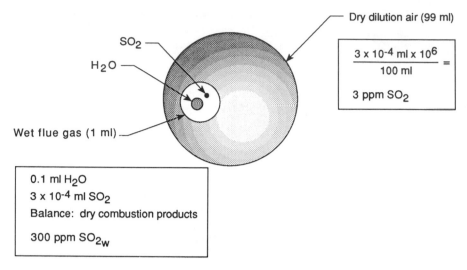

Dry dilution air (99 ml)

$SO_2$

$H_2O$

$$\frac{3 \times 10^{-4} \text{ ml} \times 10^6}{100 \text{ ml}} =$$

3 ppm $SO_2$

Wet flue gas (1 ml)

0.1 ml $H_2O$
$3 \times 10^{-4}$ ml $SO_2$
Balance: dry combustion products

300 ppm $SO_{2w}$

**FIGURE 3-17.**    Example of a wet flue gas being diluted with dry air.

concentration of $3 \times 10^{-4} \times 10^{-6} = 300$ ppm. Because, in this example, the 1 ml of the flue-gas sample also contains 0.1 ml of water vapor, the 300 ppm value is actually a wet basis value.

To report a dry basis $SO_2$ emission concentration, the moisture content would have to be known. The expression used to do this is

$$c_d = \frac{c_w}{(1 - B_{ws})} \tag{3-3}$$

where $c_d$ = dry basis concentration value (in parts per million or percent)
$\quad c_w$ = wet basis concentration value (in parts per million or percent)
$\quad B_{ws}$ = the moisture fraction of the flue gas

On the other hand, if the diluted sample were dried using a chiller or permeation dryer, the moisture content would still need to be known. Because the dilution ratio applies to the wet gas, the measured, dry concentration would have to be converted to a wet basis before the ratio could be used to scale the value up to source level:

$$c_d = \frac{c_{\text{measured, dry}}(1 - B_{ws}/R)}{(1 - B_{ws})} \tag{3-4}$$

A value for $B_{ws}$ can be obtained by either installing some type of moisture monitor, using a value obtained by manual stack testing, or by estimating a value based on process parameters. Using a moisture analyzer is the most straightforward approach; however, this adds another analyzer to the CEM system. A common technique is to measure oxygen on both a wet and a dry basis and use the results to compute $B_{ws}$. Also, various calculation approaches have been proposed (McGowan 1976, Aldina 1985) that involve the manipulation of combustion source $F$ factors. But frequently, a moisture value obtained from manual stack test measurements is set as a constant factor to make the appropriate corrections. This approach assumes that variation in the value will be small under normal source operating conditions.

It is not necessary to convert dilution measurements to a dry basis if emission measurements are to be reported on an actual, or wet, basis. If the emissions are to be reported in terms of a pollutant mass rate (pollutant mass per unit time) as in the U.S. EPA acid rain program under Title IV of the 1990 Clean Air Act Amendments, the wet basis measurement can be used directly. Also, if emissions are reported in terms of nonograms per joule or in pounds per million British thermal units, a moisture correction is not needed if $CO_2$ is monitored along with the pollutant. This is due to the form of the equation used to calculate emission rate in nanograms per joule. When $CO_2$ is monitored as a diluent and the $F_c$ factor equation is used, moisture content cancels out in the ratio between the pollutant concentration and $CO_2$ concentration, as is discussed in Chapter 8 and in U.S. EPA (1990).

## SAMPLING INTERFACE–MONITOR CALIBRATION

The entire extractive sampling system and gas analysis system must be capable of being calibrated as a unit. In the design of the system, calibration gases should be able to be injected as close as possible at the probe—a recommendation made by the U.S. EPA in its Appendix F quality assurance requirements. This is necessary to check for leaks or other operational problems in the system. The analyzer should be calibrated at the same gas flow rate, pressure, temperature, and operating procedures that are used in monitoring the stack gas. Flooding the coarse filter or dilution probe with calibration gas at the inlet or using a check valve that allows calibration gas injection directly behind the coarse filter are the best methods for performing this check. If the system is not pressurized by the calibration gas, this method can be used to determine if leaks, blockage, or gas adsorption are occurring in the system.

The system should also be capable of checking the calibration directly at the analyzer. A consideration of instrument readings from a probe calibration check and analyzer calibration check is frequently useful in system troubleshooting.

The calibration gases are usually injected automatically every 24 h, although some operators prefer to perform calibration checks manually so that the system can be watched more closely.

## EXTRACTIVE SYSTEM COMPONENTS
## AND ACCESSORIES

The design of an extractive system involves considering more than the piecing together of a probe, a pump, and some conditioning or dilution system. The task is more complex because decisions must be made on at least the minimum factors listed here:

1. Sample probe—construction and composition
2. Probe blowback—design, frequency
3. Sample line—composition, length, diameter
4. Valves and fittings—construction and composition
5. Pressure and vacuum meters—quality
6. Moisture conditioning system—refrigerated, dilution, capacity, design, construction
7. Filters—coarse, fine, coalescing
8. Pumps—capacity, type, quality
9. Cabinets or shelters—location, temperature stability
10. System controller—microprocessor to sequence and control automatic functions
11. Electrical support—fuses, circuit breakers, regulating equipment
12. Calibration gases—location, injection point, tubing regulations, gas certification

Many of these factors are discussed in further detail in McNulty et al. (1974) and Podlenski et al. (1984). The resistance of different materials to acid gases, flow rate considerations, and condensation requirements are particularly addressed. However, to construct a working extractive system that delivers a representative sample to the gas analyzers is not something that can be done without experience and an understanding of the interaction of gases, temperature, and materials. This experience is most often gained through trial and error and may require a period of time to acquire.

### References

Aldina, G. J. 1985. Continuous emissions monitoring system for dry basis pollutant mass rate measurements. Private communication.

Bergshoeff, G., and van Ijssel, F. W. 1978. Monitoring gases—A new stack sampler: Parts 1 and 2. *Internat. Environ. Safety*. March 1978:32–34 and June 1978:134–136.

Brouwers, H. J., and Verdoorn, A. J. 1990. A simple and low cost dilution system for in-situ sample conditioning of stack gases. In *Proceedings—Specialty Conference on Continuous Emission Monitoring: Present and Future Applications*. Air and Waste Management Association, Pittsburgh, pp. 380–389.

Electric Power Research Institute. 1988. *Continuous Emission Monitoring Guidelines: Update*. Report CS-5998. Electric Power Research Institute, Palo Alto, CA.

Frank. H., and Mullowney, R. 1990. Recycling ambient monitors as stack gas monitors at Dairyland Power. In *Proceedings—Specialty Conference on Continuous Emission Monitoring: Present and Future Applications*. Air and Waste Management Association pp. 390–392.

Kertzman, J. 1973. Continuous drying of process sample streams. Paper presented at the Instrument Society of America Meeting. Paper AID 73425, pp. 121–124.

Maurice, R. L., Robertson, J. A., and Howder, J. M. 1986. Design, specification, and installation of a replacement CEM at Apache Station. In *Transactions—Continuous Emission Monitoring: Advances and Issues*. Air Pollution Control Association, Pittsburgh, pp. 44–51.

McGowan, G. F. 1976. Discussion of alternative emission measurement schemes for wet scrubber applications. Technical Note, Lear Siegler, Inc.

McNulty, K. J., McCoy, J. F., Becker, J. H., Ehrenfeld, J. R., and Goldsmith, R. L. 1974. *Investigation of Extractive Sampling Interface Parameters*. EPA-650/2-74-089.

Myers, R. L. 1986. Field experiences using dilution probe techniques for continuous source emission monitoring. In *Transactions—Continuous Emission Monitoring: Advances and Issues*. Air Pollution Control Association pp. 431–439.

Podlenski, J., Peduto, E., McInnes, R., Abell, F., and Gronberg, S. 1984. *Feasibility Study for Adapting Present Combustion Source Continuous Monitoring Systems to Hazardous Waste Incinerators*, Vol. 1, *Adaptability Study and Guidelines Document*. EPA-600/8-84-011a.

U.S. Environmental Protection Agency (U.S. EPA). 1990. Reference method 19—determination of sulfur dioxide removal efficiency and particulate matter, sulfur dioxide, and nitrogen oxides emission rates. *U.S. Code of Federal Regulations*, 40 CFR 60 Appendix A.

Whitt, T. A., Defriez, H., and Collier, S. 1985. Laboratory and field evaluations of the TRS Systems, Inc., TRS-2000 total reduced sulfur CEM. In *Transactions—Continuous Emission Monitoring: Advances and Issues*. Air Pollution Control Association, Pittsburgh, pp. 114–127.

**Bibliography**

Chapman, R. L. 1974. Continuous stack monitoring, *Environ. Science Technol.* 8(6):520–525.

Federal Ministry for the Environment, Nature Conservation and Nuclear Safety. 1988. *Air Pollution Control Manual of Continuous Emission Monitoring*. Bonn, Germany.

Jahnke, J. A., and Aldina, G. J. 1979. *Continuous Air Pollution Source Monitoring Systems—Handbook*. EPA 625/6-798-005.

Laird, J. C., Patton, J. C., Zolner, W. J., and Tomlin, R. L. 1978. Unique extractive stack sampling system for continuous emission monitoring. Paper presented at the Instrument Society of America Meeting, Houston. Paper 78-87664-408-6.

Navarre, A. J., and Ayer, C. 1978. Development of Champion Papers' stack gas sampling/calibration system using a calibration probe. Paper presented at the Air Pollution Control Association Meeting, Houston. Paper 78-47.5.

Patton, J. C., Martin, G. W., Ross, J. A., and Spangler, W. 1979. A turn-key extractive sampling system to continuously monitor gaseous emissions from fossil fuel fired boiler stacks. Paper presented at the Air Pollution Control Association Meeting, Cincinnati. Paper 79-35.1.

Peritsky, M. M., Wood, R. D., and Wendt, J. O. L. 1981. Extractive flue-gas sampling challenges in-situ methods. *Power*. December 1981:48–50.

Richards, J. A. 1984. *Guidelines on Preferred Location and Design of Measurement Ports for Air Pollution Control Systems*. EPA-340/1-84-017.

Sorrel, C. B. 1986. Critical orifice in gas sampling trains for volume and rate measurements. Paper presented at the Air Pollution Control Association Meeting, Minneapolis. Paper 86-71.1.

Wolf, P. C. 1975a. Continuous stack gas monitoring part two: Gas handling components and accessories. *Pollution Engineering* 7:26–29.

Wolf, P. C. 1975b. Continuous stack gas monitoring part three: Systems design. *Pollution Engineering* 8:36–37.

Wyss, A. W., and Stroud, B. D. 1977. Design and operation of a sampling interface for continuous source monitors. Paper presented at the Air Pollution Control Association Meeting, Toronto. Paper 77-27.2.

# 4

## Introduction to the Analytical Methods

The heart of any CEM system, whether extractive or in-situ, consists of the analyzers. An extractive system transports and conditions the flue gas, but the analyzers perform the job of measurement. The selection or evaluation of CEM system analyzers must consider both regulatory specifications and performance characteristics. Although most analyzers are advertised as meeting or exceeding required specifications, care must be exercised in their selection because an analyzer's performance in the field can differ greatly from its performance on a laboratory bench.

CEM system analyzers must measure gases without interference from other gases. They are increasingly being required to measure accurately in low concentration ranges, and they must perform well in often hostile environments. Current U.S. EPA and International Standards Organization (ISO) standards for CEM system analyzers do not specify analytical techniques that are to be used (except for opacity monitors), but rather provide performance-based specifications. Therefore, it is left to the CEM system manufacturer or the user to determine the measurement techniques that would be the most appropriate for a given application. This chapter will provide a basis for the understanding of several of the techniques employed in commercially available CEM system analyzers.

## THE PROPERTIES OF LIGHT

The majority of instruments used in CEM systems are based on principles associated with the interaction of light with matter: Opacity monitors measure the effects of light scattering and absorption; a nondispersive infrared analyzer measures the amount of light absorbed by a pollutant molecule; and a chemiluminescence analyzer senses the light emitted in a chemical reaction. One could treat the operation of CEM system analyzers as "black boxes" that give out answers, but an understanding of their operation is necessary in order to apply an analyzer properly for monitoring at a source. Although some techniques will work well in a given application, others will have continuing problems in the same application. Also, for proper evaluation of system performance, a knowledge of how an analyzer works is essential in order to understand the effects of interferences, temperature, pressure, and so on. This understanding will be developed by first reviewing some of the basic properties of light and how these properties are used in the basic electro-optical analyzer.

### The Wave Nature of Light

Light can be characterized as a wave composed of oscillating electric and magnetic fields. Light waves, or electromagnetic waves as they are better termed, are distinguished by their wavelength or frequency. Figure 4-1 illustrates the typical sinusoidal nature of an electromagnetic wave and its associated wavelength.

The length between successive oscillations of a wave is called the wavelength ($\lambda$). The period of time that it takes a wave to go through an oscillation cycle is called the frequency ($\nu$). The following relationship exists between wavelength and frequency:

$$\text{frequency of light} = \frac{\text{velocity of light}}{\text{wavelength of light}}$$

$$\nu = \frac{c}{\lambda} \tag{4-1}$$

where $c$ is the velocity of light, $3.0 \times 10^8$ m/s.

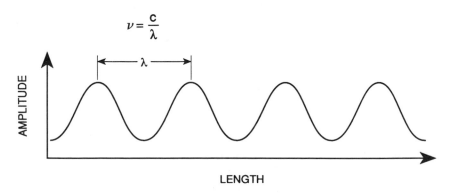

**FIGURE 4-1.**     An oscillating electric field and its wavelength.

Literature describing continuous monitoring instruments often specifies the wavelength in order to characterize the spectral region used in the analytical method. Different units for wavelength are often used for different regions of the electromagnetic spectrum, although the nanometer (1 nm = $10^{-9}$ m) has become the standard unit. Other units are, for example, the angstrom (1 Å = $10^{-10}$ m), which has been used historically in the ultraviolet region. In the infrared region, both the micrometer (1 $\mu$m = $10^{-6}$ m, also termed 1 micron) and the wave number are commonly used by spectroscopists. The wave number is expressed as

$$\bar{\nu} = \frac{1}{\lambda} \quad (\text{cm}^{-1})$$

(4-2)

Note that the units of $\bar{\nu}$ are expressed in terms of the number of wavelengths per centimeter, called reciprocal centimeters or wave numbers. The wave number $\bar{\nu}$ is essentially a measure of frequency, differing from $\nu$ by the constant factor of the velocity of light. To change between the two designations, obtain the reciprocal of the wavelength expressed in micrometers ($\mu$m) and multiply by $10^4$ to obtain wave numbers in units of cm$^{-1}$.

Light used in continuous monitoring instrumentation ranges from 200 nm in the ultraviolet to 6000 nm in the infrared. Figure 4-2 shows the regions of the electromagnetic spectrum used in analyzers incorporated in CEM systems.

### Absorption of Light by Molecules

Light carries energy. Light has the properties of a wave, but in some of its interactions with matter it behaves as if it were composed of discrete packets of energy, called photons. In a beam of light having frequency $\nu$, each of these photons carries an energy defined by the Einstein–Planck relation of Equation (4-3):

$$E = h\nu = \frac{hc}{\lambda}$$

(4-3)

where $h$ is Planck's constant and has a numerical value of $6.62 \times 10^{-27}$ erg-s. Clearly, the energy of a photon is dependent upon the frequency or wavelength of the light. Light (photons) of short wavelengths (such as in the ultraviolet) will carry more energy than light (photons) of longer wavelengths (such as in the infrared). A light beam is more or less intense depending upon whether it has more or fewer photons per unit time. Photons of different energies will do different things. From a practical

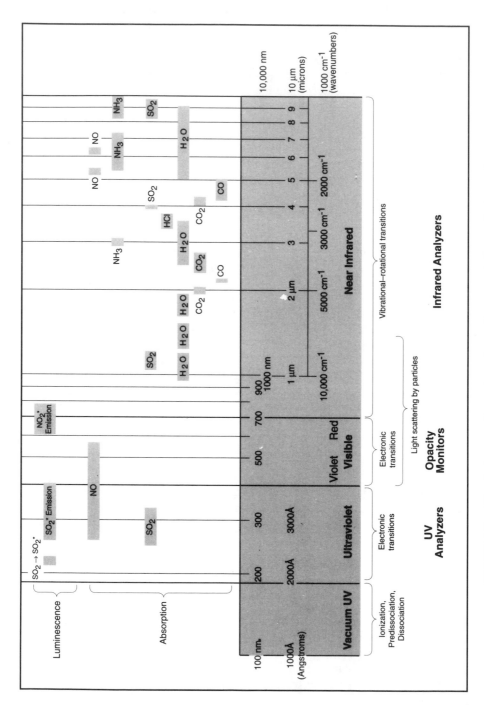

**FIGURE 4-2.** The electromagnetic spectrum for continuous emission monitoring analyzers.

sense, sitting on the beach too long in bright sunlight can cause a severe sunburn. On the other hand, sitting under an infrared heat lamp will soothe sore muscles without causing sunburn. In terms of intensity, the effects of UV radiation are extremely severe under the antarctic ozone hole, but are less severe where the stratospheric ozone layer reduces the number of photons per unit time reaching the earth's surface. Similarly, light of different wavelengths (photons of different energies) will have a variety of effects on molecules. In a monitoring instrument, the manufacturer determines the best way to use these effects to make gas concentration measurements.

Molecules are made up of atoms and molecular electrons that are arranged in very specific patterns, which undergo unique and complex motions. If light of a given wavelength should resonate with one of these motions, it will have a high probability of being absorbed by the molecule. The light essentially alters the molecular energy, causing the molecule to act differently than it was acting before the light was absorbed.

If the absorbed light is of low energy [long wavelength, low frequency ($E = h\nu$)], it will cause the molecule to undergo a specific type of rotation. This occurs typically for light wavelengths in the far infrared region of the spectrum, at wavelengths greater than 20 $\mu$m. Light in the range of 1.5–20 $\mu$m (1500–20,000 nm) can cause changes in the vibrational characteristics of a molecule. Figure 4-3 illustrates some of the specific motions that occur when photons of the right energy (light of the right wavelength) are absorbed. Figure 4-2 illustrates the regions over the range of 1–10 $\mu$m (1000–10,000 cm$^{-1}$) in the infrared spectrum, where typical pollutant and combustion gases absorb light due to vibrational–rotational transitions.

In the ultraviolet and visible regions of the spectrum, 180–700 nm, impinging light can cause the molecular electrons to change their energy states. Here, the higher-energy photons cause the electrons to become excited and in the far ultraviolet may even cause the molecule to dissociate. $SO_2$ shows a particularly strong absorption centered at 280 nm, which

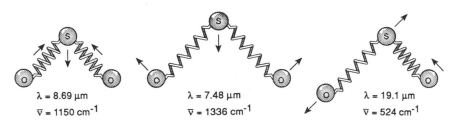

**FIGURE 4-3.**    Example of normal vibrations of the $SO_2$ molecule.

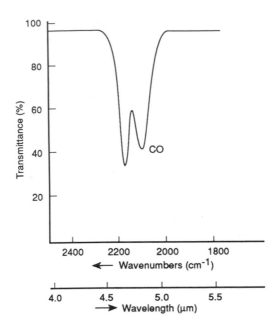

**FIGURE 4-4.**    A typical transmission spectrum.

has been taken advantage of in several $SO_2$ analyzers—as we shall see in the next chapter.

Each of these absorption processes requires a precise quantity of radiant energy. The probability of the light being absorbed and the transition occurring is greatest when $h\nu$ equals that energy. If the light passing through gas contained in a cell is changed over a the range of wavelengths, a detector located on the other side of the gas would sense a dip in the light intensity it receives at the wavelengths where these transitions occur. This is shown in Figure 4-4. This absorption can also be plotted directly in an absorption spectrum as shown in Figure 4-5. The absorption spectrum offers some advantages for quantitative analysis.

Each electronic state of a molecule will contain many vibrational energy levels and each vibrational energy level will contain many rotational energy levels. This is illustrated in Figure 4-6, which shows the possible energy states in which a molecule can exist. The molecule's energy state can be modified by supplying a photon of the proper energy that can cause a transition from one state to another. Because there are a large number of states, there will also be a large number of wavelengths at which light will be absorbed.

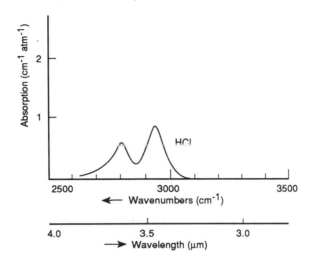

**FIGURE 4-5.**    A typical absorption spectrum.

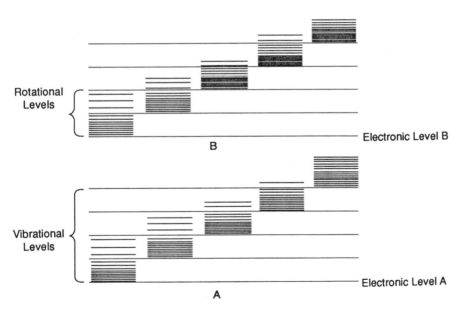

**FIGURE 4-6.**    Energy level diagram for a molecule.

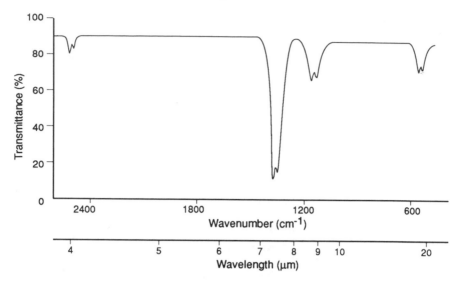

**FIGURE 4-7.**    Infrared vibrational–rotational transmission spectrum for $SO_2$.

The total energy of a molecule in a specific energy state can be summarized by the approximation.

$$E_{\text{total}} = E_{\text{electronic}} + E_{\text{vibrational}} + E_{\text{rotational}} \qquad (4\text{-}4)$$

The fact that transitions can occur between many of these states implies that energy will be absorbed at many different wavelengths. This gives rise to the absorption spectrum that typifies a molecule. As an example, Figure 4-7 illustrates the vibrational–rotational absorption spectrum of $SO_2$ in the near infrared region. Such spectra are very important in the development of analytical techniques for gas monitoring.

Also, Figure 4-2 shows that different molecules can absorb light in the same region of the spectrum. This can cause problems in developing an analyzer because it is usually difficult to distinguish the relative amounts of absorption from each compound in the overlap region. Water vapor can be particularly troublesome because it absorbs in many regions of the infrared spectrum and is usually present at percent levels in the gas stream, whereas pollutant gases are present at parts-per-million levels. This is one of the reasons why water is often removed prior to entering the analyzer. An alternative to removing interfering compounds is to a select a region of the spectrum where there is no overlap. It is easier to do this using high-resolution instruments, but such instruments are relatively expensive and may not be particularly adaptable for use in the field. Different approaches to such problems will be discussed further in following chapters.

To this point we have discussed the fact that molecules can absorb light energy. However, the question arises as to how this phenomenon can be expressed quantitatively. The answer lies in the mathematical expression known as the Beer–Lambert law.

### The Beer–Lambert Law

When studying the absorption of light by gases, the Beer–Lambert law can be used to relate the amount of light absorbed to the concentration of the pollutant gas. First, consider the system shown in Figure 4-8, composed of a light source, flue gas, and a sensor that measures the light intensity.

The Beer–Lambert law states that the transmittance of light through the medium is decreased exponentially by the product $\alpha c \ell$, or

$$\text{Tr} = \frac{I}{I_0} = e^{-\alpha c \ell} \qquad (4\text{-}5)$$

where Tr = transmittance of the light through the flue gas
$I_0$ = intensity of the light entering the gas per second
$I$ = intensity of the light leaving the flue gas per second
$\alpha$ = molecular absorption coefficient (dependent on wavelength $\lambda$)
$c$ = concentration of the pollutant
$\ell$ = distance the light beam travels through the flue gas

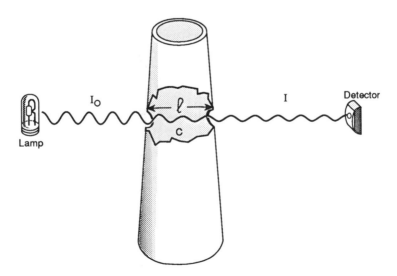

**FIGURE 4-8.**    Example of a system for measuring pollutant gas concentrations.

The absorption coefficient $\alpha$ is dependent upon the wavelength of the light and upon the properties of the pollutant molecule. The coefficient is a quantitative expression of the degree to which a molecule will absorb light energy at a given wavelength. If no absorption occurs, $\alpha$ will be zero and the transmittance would equal 100%. If an electronic or vibrational–rotational transition occurs at some wavelength, the absorption coefficient will have some value and the reduction of light energy across the path will also depend upon the pollutant concentration $c$ and the pathlength $\ell$.

Utilizing this principle, an instrument can be designed to measure pollutant gas concentrations. All that is needed is light having a wavelength that will cause some transition in the molecule of interest, a sample cell to hold the gas for measurement, and a light detector. $I_0$ is determined by taking a reading from the detector when no pollutant is in the duct or sample cell. The concentration can be obtained from the Beer–Lambert expression if $\alpha$ and $\ell$ are known, but typically, a calibration curve is generated with known gas concentrations, rather than using theoretical values for $\alpha$. Figure 4-9 shows a calibration plot typical of an instrument designed to measure light absorption by pollutant molecules.

Here, $\ln(1/Tr)$ is plotted against concentration instead of plotting transmittance against concentration. This logarithmic plot gives a straight line, from which it is much easier to develop the calibration plot. In the figure, the calibration line was developed using three calibration standards

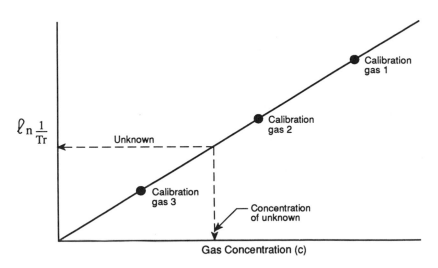

**FIGURE 4-9.**    Calibration plot for the Beer–Lambert relation.

of known concentration. The intensity $I_0$ is determined at the detector using a gas, such as air or nitrogen, containing no pollutant. Calibration gases were then injected one at a time into a sample cell to obtain values at the detector of $I_1$, $I_2$, and $I_3$. The ratios of $I/I_0$ were calculated, the logarithm taken, and values of $\ln(1/Tr)$ plotted against concentration to obtain the line. Now, injecting the unknown gas, a value for $I_u$ can be obtained. The unknown concentration can then be determined on the concentration axis by calculating $\ln(1/Tr_u)$ and locating this value on the calibration line.

Note that it was not necessary to know the value of the absorption coefficient to obtain the unknown concentration value. The use of gas calibration standards avoids that necessity in this type of instrumentation.

## LIGHT SCATTERING BY PARTICLES

There is another way to remove energy from a beam of light, other than by absorption from individual molecules. If particulate matter, aerosols, or droplets are present in the flue gas, light incident on these materials can be scattered in various directions. As a result, not all of the light will continue in its initial direction and its transmittance through the gas will be reduced. The mechanisms of light scattering are dependent upon both the size of a particle (as defined by its radius $r$) and the wavelength of light that impinges on it—different things happen depending on how $r$ compares to $\lambda$.

There are three basic types of scattering processes that occur. If the particle is very much smaller than the wavelength of light, $r/\lambda \leq 1$, then the particle–light interaction can be described by Rayleigh scattering. If the size of the particle and the light wavelength are comparable, $r/\lambda \approx 1$, then the interaction can be described by Mie scattering. When particles are very much larger than the light wavelength, $r/\lambda \geq 1$, the concepts of geometric optics (such as reflection and refraction) can be used to explain how the light scatters (macroscopic scattering).

In a typical flue gas, particle sizes may range from 0.1 to 10 $\mu$m or greater. When visible light ranging from 400 to 600 nm is directed through the gas, all of these scattering processes may take place. A description of each scattering process follows.

### Rayleigh Scattering: $r/\lambda \leq 1$

Particles smaller than about 0.1 $\mu$m will scatter visible light by Rayleigh scattering. In this case, the electric field of the light interacts with the electrons within the molecules of the particle.

The electrons are correspondingly accelerated in their motion in the molecule. It is a phenomenon of nature that an accelerated electron will emit electromagnetic radiation in all directions. The net effect is that the oscillating electron scatters light out of the light beam. As a result of this phenomenon, small particles ($< 0.1$ $\mu$m) are very effective in scattering visible light.

### Mie Scattering: $r/\lambda = 1$

As the particle size increases relative to the wavelength, the molecular electrons throughout the particle no longer see a uniform field. The intensity and field direction vary at different points within the particle and the electrons will accelerate in different directions to generate a complex scattering pattern. In fact, the scattered light waves can add together constructively or can subtract destructively as they come from the separate locations within the particle, as shown in Figure 4-10.

The size range of pollutant particles emitted to the atmosphere often corresponds to the wavelengths of visible light (400–700 nm). Bag houses and electrostatic precipitators used to control the emission of particulate matter will effectively collect particles that are greater than 1 $\mu$m (1000 nm) in diameter. It is more difficult, however, to collect particles in the submicron range ($< 1$ $\mu$m). These are the particles that will have a higher probability of escaping into the atmosphere. Light scattering from particles in flue gases is therefore typically due to Mie scattering.

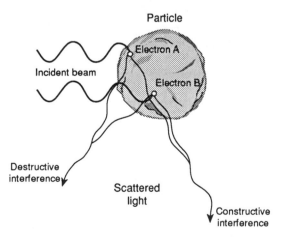

**FIGURE 4-10.**    Mie scattering: destructive and constructive interference of light waves scattered from molecular electrons when particle radius and light wavelength are comparable.

### Geometric Optics: $r/\lambda \geq 1$

Particles in a size range of 20 or more times the wavelength can be described in the particle–light interaction by using concepts of geometrical optics. The incident beam of light can be thought of as being composed of separate rays that interact with the particle. This interaction can be in the form of the processes of reflection, refraction, or diffraction (Figure 4-11).

Reflection is a surface effect where, without entering the particle, the light ray changes direction after striking it. Reflection will occur if the depth of particle surface irregularities are small relative to the wavelength of the light. Light refraction occurs after light enters the particle. Its speed and direction change because of the change of the optical characteristics (refractive index) of the material. Once light has entered the particle, it can also reflect internally. Diffraction is the bending of light around an object caused by the interference effects at the particle surface.

Scattering phenomena have been emphasized here, but it should be noted that particles can also *absorb* light energy, as do gaseous molecules. When light of a specific wavelength resonates with the molecules that constitute the particle, it will cause changes from one molecular energy state to another to occur. As stated earlier, only light wavelengths that correspond exactly to the molecular energy levels will be absorbed to impart energy to the molecule. As a result, light is again lost from the incident beam.

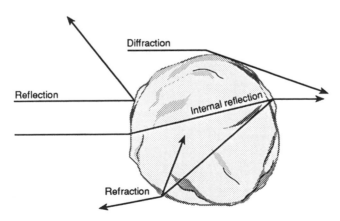

**FIGURE 4-11.**    Scattering of light from large particles ($r/\lambda \geq 1$), where geometrical optics apply.

## Opacity and Transmittance

The transmission of light through a flue gas that contains particulate matter will be reduced through a combination of scattering and absorption processes. The scattering and absorption of light by a stack plume gives rise to its opacity—its opaqueness to the transmission of light.

If light is not able to penetrate through a plume, the plume is said to be opaque—the opacity of the plume is 100%. Transmittance and opacity can be related in the following manner:

$$\text{transmittance } (\%) = 100 - [\text{opacity } (\%)]$$

Therefore, if a plume or object is 100% opaque, the transmittance of light through it is zero. If it is not opaque, but has 0% opacity, the transmittance of light correspondingly would be 100%. A stack gas or plume rarely will have either 0 or 100% opacity, but will have some intermediate value. Opacity standards are typically set at 20%, but can range from 0 to 35% opacity.

## Bouguer's Law

Particle scattering and absorption effects can be treated in a manner similar to that given in the previous discussion on light absorption by gas molecules. In an expression similar to the Beer–Lambert law, the expression called Bouguer's law is used when discussing particle scattering. Bouguer's law states that transmittance is decreased exponentially by the product $naQ\ell$ as shown in Equation (4-6):

$$\text{Tr} = e^{-naQ\ell} \tag{4-6}$$

where $n$ = number concentration of the particles
$\quad\quad a$ = projected area of the particles
$\quad\quad Q$ = particle extinction coefficient
$\quad\quad \ell$ = length of the light path through the flue gas

The expression was first stated by P. Bouguer in 1760 and was later rediscovered by Lambert. In deference to the earlier work of Bouguer, it has become customary in the field of opacity monitoring to call it Bouguer's law (Middleton 1968).

The parameter $Q$ is the particle extinction coefficient, which is dependent on the wavelength and expresses how the particle of size $a$ scatters and absorbs light of wavelength $\lambda$. It is analogous to the absorption coefficient $\alpha$, discussed previously. This expression is overly simplified, because in typical flue gas a wide range of particle sizes are present and the light used in opacity monitoring instruments usually extends over a range from 400 to 600 nm. [See Jahnke (1984) for more-detailed expressions].

Bouguer's law is an exponential relationship between transmittance and particulate matter concentration and is somewhat difficult to use in stack emission calculations. Another expression, called optical density, is frequently used in opacity monitoring and is related to opacity in the following manner:

$$\text{(optical density)} = D = \log_{10} \frac{1}{1 - \text{opacity}} = \log_{10} \frac{1}{\text{Tr}} \quad \text{(4-7)}$$

Optical density is useful in emission calculations, because it is related directly to particulate concentration, as shown in the following:

$$\ln \text{Tr} = -naQ\ell \quad \ln \frac{1}{\text{Tr}} = naQ\ell = 2.303 \, D \quad \text{(4-8)}$$

and, therefore,

$$D = \frac{naQ\ell}{2.303} \quad \text{(4-9)}$$

In terms of particulate concentration $c$ instead of the particle number density $n$, the expression can be written

$$D = \frac{A_E c \ell}{2.303} \quad \text{(4-10)}$$

where $A_E = \pi r^2 Q / m$, the specific mass extinction coefficient
$r$ = radius of the particle
$m$ = particle mass.

This expression merely states that optical density is directly proportional to the particulate matter concentration and also to pathlength. It is a very useful expression, because if, for example, the pathlength should increase by a factor of 2, the optical density will increase by the same factor. If the particulate matter concentration is decreased by $\frac{1}{2}$, $D$ decreases by $\frac{1}{2}$ also. Applications of this expression will be examined in Chapter 7 on opacity monitors.

## COMPONENTS OF AN ELECTRO-OPTICAL
## ANALYZER—BUILDING AN INSTRUMENT

The types of analytical technique used in today's commercially available CEM monitoring analyzers were listed in Table 1-1. Except for the electroanalytical methods, the analyzers generally incorporate four essential components:

1. Radiation sources
2. Spectral limiters
3. Detectors
4. Optical components

These components differ depending upon the region of the spectrum in which the instrument operates and on the analytical technique itself. The following sections give examples of these components.

### Radiation Sources

Heated materials emit radiation in the infrared region of the spectrum. Among those used are Nernst Globars (fused hollow rods of zirconium and yttrium oxides, heated to about 1500°C), Globars (heated rods of silicon carbide), carbon rods, and heated nichrome wire. A newer and more sophisticated source is the diode laser, a solid-state infrared source that has only recently begun to be incorporated in monitoring instruments after its introduction for this use in 1971 (Imasaka and Ishibashi 1990; Hinkley and Kelley 1971).

Visible light is used principally in opacity monitors, where the peak spectral response is required to be between 500 and 600 nm. Suitably filtered tungsten lamps, quartz–halogen lamps, and green-light-emitting diodes have been used for this application.

For the ultraviolet region of the spectrum, hollow cathode gas discharge tubes, high-pressure hydrogen or deuterium discharge lamps, xenon arc lamps, and mercury discharge lamps have been used. However, problems do exist in obtaining stable UV intensities over the extended periods of time required for the lamp operation. Instrument manufacturers pride themselves on the techniques that they develop to solve such problems.

### Devices that Isolate Limited Spectral Regions

Spectral limiters are used to restrict the wavelengths of light to those that cause the changes in the molecular energy levels of the pollutant molecule being measured. The simplest example is a filter that allows only light in a narrow spectral region to pass through it. Interference filters, constructed by vacuum deposition of metallic films onto glass or other materials, are commonly used in the infrared region of the spectrum. Coated, front-surfaced mirrors can be used in the UV region for wavelength selection.

Diffraction gratings are commonly used in the ultraviolet and visible regions of the spectrum to select specific wavelengths. A grating consists of a reflective surface finely ruled or etched with a large number of parallel lines (on the order of 600 lines/mm). Each line will scatter light impinging on it, causing the light to interfere constructively and destructively to spread out a reflected spectrum.

## Detectors

The type of detector used in an analyzer is very dependent on the energy of the light that it is sensing. Because light in the infrared region is relatively weak in the energy it carries, special stratagems are often devised to detect infrared intensity changes. Pneumatic, microphone-type detectors (Luft detectors) traditionally have been used in pollutant analyzers; however, solid-state detectors, cooled with Peltier coolers, are becoming more common. Sensitivity is often increased by not overly limiting the spectral region of the analyzer, but using a broader band of radiation to obtain more light for the detector. The special methods of gas filter correlation and Fourier transform infrared spectroscopy take advantage of this technique.

The most familiar detector in the visible region is the human eye, which is of course used as the detector in EPA Reference Method 9. Phototubes, photomultiplier tubes, and photovoltaic cells are used in instrumented systems. More recently, photodiode arrays are being incorporated into CEM analyzers (Saltzman 1990; Durham et al. 1990). The diode arrays provide a simple way of measuring multiple wavelengths and have great potential for the development of multigas analyzers.

## Optical Components

In most electro-optical analyzers, additional components are often used to direct and focus light. Rotating shutters are used to reflect or block light to create oscillating signals. Half-silvered mirrors will both reflect and allow light to pass through them—an extremely useful technique, especially in opacity monitors. Lenses, slits, and diaphragms are used to focus light in the system, and plain glass windows are used to separate the primary optical system from flue gases.

These components are used in various combinations in the construction of pollutant gas and opacity monitors. These combinations will be examined in further chapters that discuss the many unique systems that are commercially available today.

## References

Durham, M. D., Ebner, T. G., Burkhardt, M. R., and Sagan, F. J. 1990. Development of an ammonia slip monitor for process control of $NH_3$ based $NO_x$ control technologies. In *Proceedings—Speciality Conference on Continuous Emission Monitoring: Present and Future Applications*. Air and Waste Management Association, Pittsburgh, pp. 298–313.

Hinkley, E. D., and Kelley, P. L. 1971. Detection of air pollutants with tunable diode lasers. *Science* 171:635–639.

Imasaka, T., and Ishibashi, N. 1990. Diode lasers and practical trace analysis. *Anal. Chem.* 62(6):363A–371A.

Jahnke, J. A. 1984. *Transmissometer Systems—Operation and Maintenance, An Advanced Course*. EPA 450/2-84-004.

Middleton, W. E. K. 1968. *Vision Through the Atmosphere*. University of Toronto Press, Toronto.

Saltzman, R. S. 1990. A process UV/VIS diode array analyzer for source monitoring. In *Proceedings—Specialty Conference on Continuous Emission Monitoring: Present and Future Applications*. Air and Waste Management Association, Pittsburgh, pp. 227–238.

## Bibliography

Barrow, G. M. 1962. *Molecular Spectroscopy*. McGraw-Hill, New York.

Jahnke, J. A., and Aldina, G. J. 1979. *Continuous Air Pollution Source Monitoring Systems—Handbook*. EPA 625/6-79-005.

Willard, H. H., Merritt, L. L., and Dean, J. A. 1987, *Instrumental Methods of Analysis*. Van Nostrand, New York.

# 5

## Extractive System Analyzers

The choice of gas analyzers for an extractive CEM system is important because some analytical techniques are more appropriate than others in a source application or system design. The total system must be considered; any system evaluation must be made on both the analyzers used and the extractive conditioning system itself. Chapter 3 reviewed methods for extracting, transporting, and conditioning flue gases for analysis. This chapter discusses the types of instrumentation used to determine gas concentration measurements, after the flue gas has been properly conditioned for analysis.

Many problems occur when monitoring emissions from stationary sources, one of the major problems being that the pollutant gas must be analyzed in the midst of other stack gas constituents. The possibility of these other gases interfering in the measurement process is great, and a number of approaches have been taken to avoid such interferences. This has led to the application of a variety of existing measurement techniques and to the development of several new methods unique to emissions monitoring.

In Table 2-1, a summary was given of the current commercially marketed techniques used in source monitoring analyzers. Table 5-1 gives the major categories of methods used in extractive system analyzers. Extractive analyzers utilize techniques from all of these categories, whereas in-situ monitors use only spectroscopic absorption or electrochemical methods. Extractive analyzers are designed to monitor at source-level gas concentrations or at ambient pollutant concentrations when used in conjunction with a dilution system. It is important to understand the basic operating principles of these analyzers in order to evaluate properly the

77

**TABLE 5-1    Analytical Methods Used in Extractive System Gas Analyzers**

| Absorption Spectroscopic Methods | Luminescence Methods | Electroanalytical Methods | Paramagnetic Methods |
|---|---|---|---|
| Infrared Ultraviolet | Fluorescence Chemiluminescence Flame photometry | Polarography Potentiometry Electrocatalysis | Thermomagnetic Magnetodynamic Magnetopneumatic |

appropriateness of the technique in a system, the effects of interferences, and the levels of instrument maintenance that will be required.

## SPECTROSCOPIC METHODS—INFRARED MONITORING INSTRUMENTS

Infrared analyzers have been developed to monitor heteroatomic molecules such as $SO_2$, NO, CO, HCl, and $CO_2$, as well as hydrocarbons and other gases. Nondispersive techniques that use optical filters or gas filters have been used in source monitoring applications for over 20 years and are well established. The technique of Fourier transform infrared (FTIR) spectroscopy has been used primarily in research and analytical laboratories, but is now being used increasingly in remote sensing and source monitoring applications.

### Nondispersive Infrared Analyzers

Instruments designed for measuring molecular light absorption can be classified as spectrometers or spectrophotometers. Spectrometers are designed to vary the wavelength over a range of wavelengths to obtain the type of detailed wavelength-dependent absorption information shown in Figure 4-5. Spectrophotometers, on the other hand, do not disperse light to scan the spectrum, but use filters or some other mechanism to measure light absorption over a relatively small range of wavelengths, or 'bands," centered at an absorption peak of the molecule. Spectrophotometers used in the infrared region are commonly called nondispersive infrared (NDIR) analyzers. Because of their simplicity, NDIR analyzers are frequently used in field applications.

In Chapter 4, Figure 4-2 illustrated the location of principal infrared absorption bands for commonly monitored gases. Note from the figure that a number of the gases have absorption bands that occur in the same

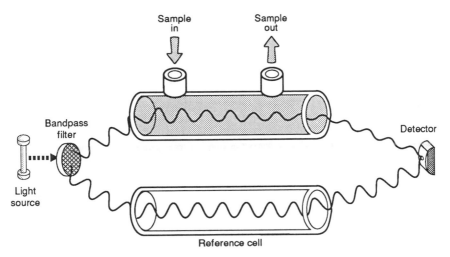

**FIGURE 5-1.**    A simple nondispersive infrared (NDIR) analyzer.

region of the spectrum. In particular, water and $CO_2$ have broad bands that can interfere in the measurement of several important pollutant species. The CEM system designer must choose an analyzer that will measure in a region where such an interference does not occur, use another technique that is not sensitive to these gases, or else remove the interfering gases.

In a simple NDIR analyzer (Figure 5-1), infrared light is emitted from a radiation source such as a Nernst glower, a Globar, a heated nichrome coil, or other type of infrared radiator. The light is transmitted through two gas cells: a reference cell and a sample cell.

The reference cell contains a gas, such as nitrogen or argon, that does not absorb light at the wavelength used in the instrument. As the transmitted beam passes through the sample cell, pollutant molecules will absorb some of the infrared light. As a result, when the light emerges from the end of the sample cell, it will have less energy than when it entered. It also will have less energy than the light emerging from the reference cell. The energy difference is sensed by some type of detector, such as a solid state sensor [e.g., mercury cadmium telluride (MCT), lead sulfide (PbS), arsenic triselenide $(As_2Se_3)$] or a pneumatic-type sensor. The sensitivity of solid-state sensors may be increased by cooling with a thermoelectric chiller. Thermistor- and thermocouple-type detectors are usually not sensitive enough to measure the small energy differences that result.

The ratio of the detector signals from the two cells gives the light transmittance $Tr = I/I_0$, which can be related to the pollutant gas concentration [see Equation (4-5) and Figure 4-9]. In the reference cell, where $c_{\text{reference}} = 0$, a value for $I_0$ can be easily obtained, because:

$$I_{\text{sample}} = I_0 e^{-\alpha c \ell} \qquad (5\text{-}1)$$

$$I_{\text{reference}} = i_0 e^{-\alpha c \ell} = I_0 e^0 = I_0 \qquad (5\text{-}2)$$

The value of $I_0$ remains constant, and $I_{\text{sample}}$ can be related to it to obtain the transmittance ($\text{Tr} = I/I_0$) and a concentration measurement.

A common problem with analyzers that use a detecting arrangement as shown in Figure 5-1 is that the gases that absorb light in the same spectral region as the pollutant molecule cause a positive interference in the measurement. For example, water vapor and $CO_2$ interfere in the measurement of CO (see Figure 4-2). When this happens, these gases must be removed by some scrubbing system before the sample gas enters the analyzer. Alternatively, sample gas scrubbed of the pollutant can be passed through the reference cell. In such a flowing reference cell, the level of interference will be the same as in the sample cell, but it will also be referenced to a zero level of the pollutant being measured.

Another solution to this problem is to use a pneumatic detector with absorption chambers arranged in series, as shown in Figure 5-2. In this detector, two chambers are filled with gas of the species being measured. The front chamber is shorter than the rear chamber and both are connected to a sensor that can detect differences in gas pressure between the two cells or the flow of gas between the two cells.

To understand the operation of this detector, first consider that the infrared radiation is absorbed over a range of wavelengths corresponding to an absorption band such as that shown in the figure. Because the wavelength-dependent absorption coefficients are greater at the band center, the energy absorbed in the front chamber will be greater at the band center than at the wings. As a result, less energy at these center wavelengths will be transmitted to the rear chamber. Absorption in the second chamber will be primarily at wavelengths at the wings of the band. The absorption is weaker at the wings ($\alpha_{\text{wings}} < \alpha_{\text{center}}$), so the cell is made longer to allow for more overall absorption. Therefore, when no pollutant gas is present in the sample cell, the radiation absorbed in each cell will be approximately equivalent.

In the infrared spectral region, molecules absorbing the radiation are excited to higher energy vibrational–rotational states, which means that they will vibrate and rotate more vigorously. However, the excited molecules soon collide with other molecules and this extra energy is transferred in the collision. The energy released causes the molecules to

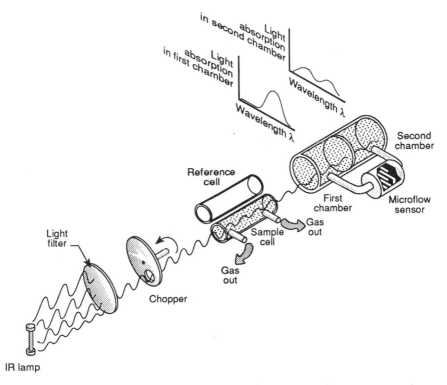

**FIGURE 5-2.**    Operation of an NDIR analyzer using a pneumatic-type sensor; series arrangement of chambers in the detector.

move faster, or heat up, increasing the temperature and pressure of the gas in the cell. If both chambers absorb the same amount of radiation, the pressures will be the same.

Now, if the pollutant being measured is present in the sample cell, there will be a loss of light energy before it reaches the pneumatic detector. The loss of energy will be primarily at the band center and the pressure in the first chamber will not be as great as before, when pollutant gas was not present in the sample cell. There will be more absorption in the rear chamber and a pressure difference will develop. This pressure difference can be detected by using either a capacitance-type microphone detector or a flow sensor. In either case, an electrical signal is developed that can be related to the pollutant gas concentration.

If an interfering gas is present in the sample, it will generally exhibit a broadband interference. If the absorption is sufficiently broad over the absorption band wavelengths of the pollutant, the degree of absorption due to the interferant will be approximately the same in the front and rear chambers. It will therefore not affect the pollutant measurement. This technique has been utilized to develop CO analyzers that minimize the effects of $CO_2$ and $H_2O$ interferences.

With regard to the types of sensors used in pneumatic detectors, the microflow sensor has been widely applied. The microflow sensor consists of a thin film grid of heated nickel that is cooled as the gas flows past the grid, thereby changing the grid's electrical resistance (Ueda and Watanabe 1978). This change is correlated to the change in pressures in the chambers and therefore to the concentration of the pollutant gas in the sample cell. The microflow sensor technique is similar to that used in thermal sensors applied in flue-gas velocity monitoring.

In contrast, the microphone-type detector is an older technology that correlates the change of capacitance between a metal diaphragm and a fixed plate to the pressure differences in each chamber. One of the problems associated with the diaphragm-type detector has been its susceptibility to vibration. Vibration from plant equipment can be transferred to the instrument, resulting in a noisy and sometimes unintelligible signal.

Other instrument configurations have been developed using pneumatic detectors. For example, NDIR analyzers have been designed using only a measurement cell, the $I_0$ reference value being obtained when zeroing the instrument with gases. Multigas analyzers have been developed by stacking detector chambers, each filled with a gas to be measured.

### Gas Filter Correlation Analyzers

A different type of NDIR technique common to in-situ monitors has recently been applied to extractive system analyzers. This technique, gas filter correlation (GFC), uses a reference cell that contains a 100% concentration of the pollutant, instead of the 0% concentration in the analyzers discussed previously. This method is currently being used to measure $SO_2$, NO, $CO_2$, CO, $NH_3$, $H_2O$, HCl, and hydrocarbons in extractive system analyzers. The essential features of a GFC instrument are shown in Figure 5-3.

In a GFC analyzer designed to measure CO, radiation from an infrared source passes through a filter wheel that contains a neutral gas, such as $N_2$, on one side and CO on the other. The light is then passed through a modulator that creates an alternating signal. The light enters the sample cell and reflects from the mirrors to make multiple passes through the gas. This increases the path length $\ell$ of the system and improves the instrument sensitivity. A portion of the light will be absorbed by any CO

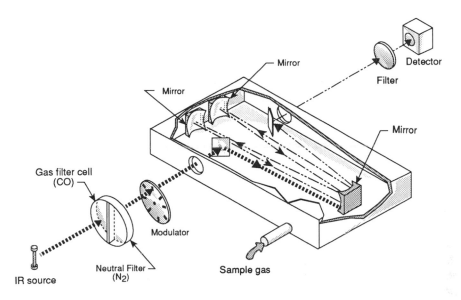

**FIGURE 5-3.**    An extractive system gas filter correlation (GFC) analyzer for monitoring CO.

molecules in the sample cell and the radiation then exits to be measured at the IR detector.

In the operation of the instrument, the filter wheel is continuously rotating. First, consider the difference in signals between the CO and $N_2$ sides of the wheel when the sample cell is flushed with zero gas. When light passes through the gas filter (the reference beam) it will be attenuated (Figure 5-4a).

The gas filter contains enough CO to remove most of the light at the wavelengths where CO absorbs. The gas filter transmits light whose intensity cannot be further reduced by CO absorption. The wavelengths not absorbed by CO are not removed and are passed on to the detector. The net result is a reduction in light energy reaching the detector. Now, when the radiation from the infrared source passes through the $N_2$ side of the wheel, the intensity is not reduced. However, in practice, a neutral density filter (an optical filter that reduces the light intensity uniformly over the wavelength range) is placed over the $N_2$ side of the wheel. This filter attenuates the signal to the same level that the gas filter side is attenuated due to CO absorption (Figure 5-4b).

Next, consider the condition where sample gas containing CO is introduced into the sample cell. The infrared beam will be alternating through

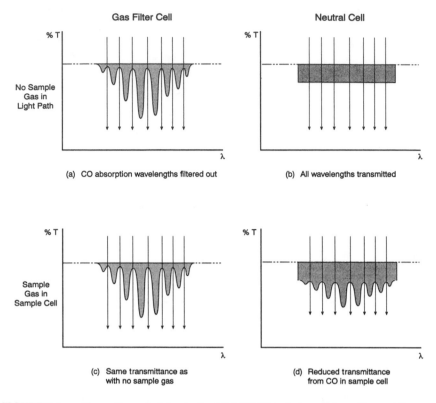

**FIGURE 5-4.**    Absorption principles in the GFC NDIR technique: (a) gas filter cell beam —no sample; (b) neutral beam—no sample; (c) gas filter cell beam—CO in sample cell; (d) neutral beam—CO in sample cell.

the $N_2$ and CO sides of the filter wheel and then will pass through the sample cell on to the detector. CO molecules are now present in the sample cell and they will absorb light energy at the CO absorption wavelengths. Because the gas filter was chosen to absorb energy at these same wavelengths, the absorption is already complete in the gas filter cell beam, and the detector will see the same signal as it did when the sample cell contained zero gas (Figure 5-4c). The beam passing through the $N_2$ side, however, will carry less energy because light is absorbed by the CO in the sample gas (Figure 5-4d). The difference in energy between the two beams is monitored and can be related to the CO concentration in the sample.

Other gases having spectral patterns in the same region as CO will not affect the measurement if they do not "correlate" with the CO spectral pattern. If, however, the peaks coincide with spectra of the CO, a positive interference will result. The interfering gas will reduce the light intensity when light passes through the $N_2$, but not when passing through the gas filter. Conversely, if the interfering molecule's peaks fall between the troughs of the CO spectra, a negative interference will result. The energy of the light passing through both the gas filter cell and neutral cell is now reduced. Because the energy through the gas filter is supposed to remain the same to use as a reference against the measurement beam ($N_2$ side), the interference will be negative.

In most cases, the interfering spectra will have some peaks that overlap and others that fall in the troughs of the CO spectra. As a result of this, positive and negative correlations will cancel to minimize the effect of the interference. Interference is further minimized by using a band-pass filter to select a region of the spectrum where the CO absorption is most specific. However, $H_2O$ and $CO_2$ interference can still be a problem in some instruments.

The GFC technique offers several advantages over the more traditional zero-gas reference cell methods. Because the method is not limited to one absorption peak, but can measure over a range of wavelengths, more energy will fall on the detector. Also, because a ratio is obtained between a measurement beam and a reference beam, variations in light intensity, resulting from infrared source fluctuations or dirt accumulation on the optics, will cancel out. For this reason, the GFC technique can be readily applied to measuring pollutant gases directly in the stack—an application that will be discussed later.

The GFC technique can be used to design instrument systems that monitor several gases at the same time. By inserting gas filters corresponding to the gases desired to be monitored, up to eight different gases have been measured in a single instrument (Dillehay 1990). Ideally, a multicomponent analyzer should reduce system costs by reducing the number of analyzers that are purchased and that need to be maintained.

## Fourier Transform Infrared Spectroscopy

The technique of Fourier transform infrared (FTIR) spectroscopy has recently been applied to source monitoring. Although traditionally a laboratory technique, the advent of relatively inexpensive and powerful microprocessor systems has allowed the method to be applied in the field. One advantage of the method is that several gases can be monitored at the same time by a single instrument. In principle, 15 or more gases, including organic species, can be monitored. This feature can find application in the monitoring of combustion sources, toxic waste incinerators, and industrial

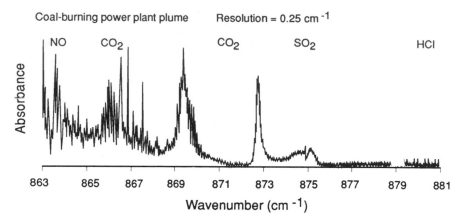

FIGURE 5-5.    Infrared absorption spectrum of a combustion gas sample (Herget, 1979).

processes. For example, the FTIR technique has been used to analyze the complex spectra of combustion gas samples such as that shown in Figure 5-5.

Basically, the FTIR technique gives a "picture" (such as Figure 5-5) of the total absorption spectrum of the sample gas over a broad spectral range—instruments typically have a range of from 2.5 to 25 $\mu$m [4000 to 400 cm$^{-1}$ (wave numbers)]. However, the camera used to take the picture is an unusual one. A spectrum like that shown in Figure 5-5 is not obtained initially; instead, a picture, called an interferogram, such as that shown in Figure 5-6 is obtained. Now, this picture is somewhat worse than a negative; not very much can be told about the sample by looking at the interferogram. Nevertheless, it contains practically all the information

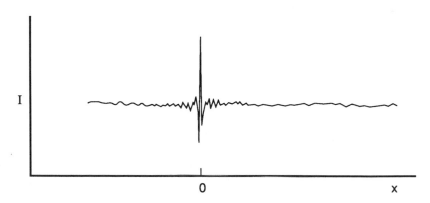

FIGURE 5-6.    A typical interferogram obtained by an FTIR spectrometer (Strong, 1979).

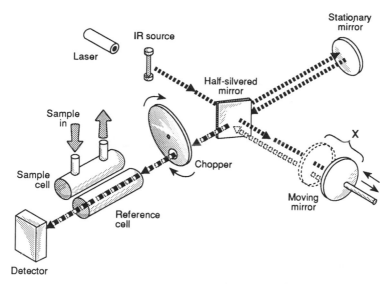

**FIGURE 5-7.**    Schematic diagram of the basic FTIR spectrometer: reference cell measurement.

about the sample that is needed. This is where the Fourier transform comes in. The Fourier transform is an elegant mathematical technique that "develops" or "transforms" the interferogram into a frequency-based spectrum such as that shown in Figure 5-5.

A microprocessor does the developing by applying the Fourier transform to the equations representing the interferogram. In addition, the system usually contains a library of spectra of the different gaseous species that are expected to be found in the sample. By comparing these library spectra to the sample spectrum, both qualitative and quantitative sample data can be obtained.

The "camera" in this method is based on the interferometer devised by Michelson in 1891. This interferometer divides the light emitted from a radiation source in two, making two beams that traverse different paths and then recombine. Figure 5-7 shows a schematic diagram of an FTIR spectrometer.

The system contains a moving mirror that can change the distance over which one of the beams travels. The helium–neon laser is used to determine the distance $(x)$ of travel of this mirror. Light from the infrared source passes through a beam splitter that partially transmits and partially reflects the light. The transmitted light is directed to the moving mirror, goes back to the beam splitter, and this time is reflected by it to the

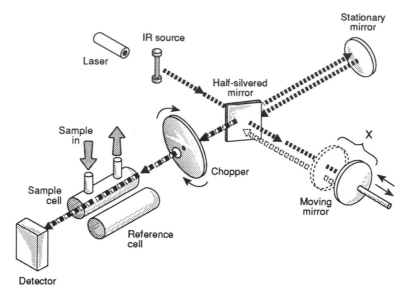

**FIGURE 5-8.**    Schematic diagram of the basic FTIR spectrometer: sample cell measurement.

detector. The light that is reflected initially from the beam splitter hits a fixed mirror, is sent back to the beam splitter, and is then transmitted to the laser detector. Using a chopper, the combined beam is alternately sent through the reference cell (Figure 5-7) and the sample cell (Figure 5-8) before it reaches the detector.

As the beams split, travel different distances, and recombine, they will constructively and destructively interfere, depending on whether they are in phase or out of phase. This interference yields the spectral information that is used to determine gas concentrations.

For a polychromatic light source in an interferometer, the interference patterns that result can be illustrated as shown in Figures 5-9 and 5-10. Figure 5-9 illustrates the interference pattern seen for the reference cell, where no sample gas is present. The pattern is given as a function of wave number for a fixed value of $x$, that is, for one position of the moving mirror. Note that the pattern is sinusoidal and is modulated by the intensity of the light at each wavelength. The total, integrated intensity of the light of Figure 5-9, $I_0(x)$, corresponds to one point on an interferogram. At each value of $x$, the moving mirror position, there will be a different pattern and a different $I_0$ value. The total reference cell interferogram is obtained by plotting similar $I_0$ values for the other values of $x$.

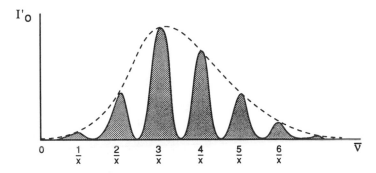

**FIGURE 5-9.** Reference cell interference pattern: intensity at the detector as a function of wave number at fixed $x$ for a polychromatic light source. (Dashed line represents the intensity distribution of the light source (Strong, 1979).

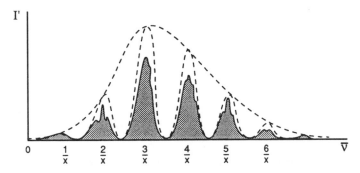

**FIGURE 5-10.** Sample cell interference pattern: intensity at the detector as a function of wave number at fixed $x$, with sample gas present. Note reduced transmittance due to absorption (Strong, 1979).

This analytical method is based on the Beer–Lambert law, as are the other infrared methods discussed in this chapter. When the infrared beam is passed through the sample cell, light will be absorbed by the gases in the cell, resulting in a reduction of the peaks shown in Figure 5-9. Similarly, an $I(x)$ value is obtained from Figure 5-10 for the sample cell interferogram.

The integrated area under the peaks in Figure 5-9 can be expressed mathematically as

$$I_0(x) = \frac{1}{2}\int_{\bar{\nu}_1}^{\bar{\nu}_2} I_0'(\bar{\nu})(1 + \cos 2\pi x\bar{\nu})\, d\bar{\nu} \tag{5-3}$$

$$I_0'(\bar{\nu}) = \frac{dI_0(\bar{\nu})}{d\bar{\nu}} \tag{5-4}$$

Similarly, the integrated area under the peaks in Figure 5-10 can be expressed as

$$I(x) = \tfrac{1}{2} \int_{\bar{\nu}_1}^{\bar{\nu}_2} I_0'(\bar{\nu}) 10^{-\alpha(\bar{\nu})c\ell}(1 + \cos 2\pi\bar{\nu}) \, d\bar{\nu} \qquad (5\text{-}5)$$

Note that the gas concentration $c$ is part of the Beer–Lambert expression incorporated in $I(x)$. However, a problem arises here. The integrals are given as a function of wave numbers, but the interferometer does not scan wave numbers, frequency, or wavelength. The interferometer simply measures the light intensity at the detector as the mirror moves (i.e., for each value of $x$). But now, the Fourier transform can be used to convert the interferogram (which gives the measured light intensity as a function of $x$) into a frequency spectrum (which gives the intensity as a function of $\bar{\nu}$).

When applied, the Fourier transform converts the integrals by switching $\bar{\nu}$s and $x$s to obtain

$$I_0(\bar{\nu}) = \tfrac{1}{2} \int_{x_1}^{x_2} \left[ I_0(x) - \tfrac{1}{2}I_0(0) \right] \cos 2\pi\bar{\nu}x \, dx \qquad (5\text{-}6)$$

and

$$I(\bar{\nu}) 10^{-\alpha(\bar{\nu})c\ell} = \tfrac{1}{2} \int_{x_1}^{x_2} \left[ I(x) - \tfrac{1}{2}I(0) \right] \cos 2\pi\bar{\nu}x \, dx \qquad (5\text{-}7)$$

which is the absorption spectrum when $I(\bar{\nu})$ is determined over a range of wave numbers.

If the two are divided, the Beer–Lambert formalism for the concentration can be brought out of the integral:

$$10^{-\alpha(\bar{\nu})c\ell} = \frac{\displaystyle\int_{x_1}^{x_2} \left[ I(x) - \tfrac{1}{2}I(0) \right] \cos 2\pi\bar{\nu}x \, dx}{\displaystyle\int_{x_1}^{x_2} \left[ I_0(x) - \tfrac{1}{2}I_0(0) \right] \cos 2\pi\bar{\nu}x \, dx} \qquad (5\text{-}8)$$

Now, by knowing $x$, by using the reference laser, and by detecting the intensities at each $x$ as the infrared beam is passed through the reference cell, $\alpha c\ell$ can be determined.

This is where the power of modern microprocessors is important. It takes numerous computations to perform the integrations and to match the sample data to normalized spectra stored in memory. A variety of algorithms have been developed to obtain concentration values from this wealth of data.

The FTIR technique is relatively new to source monitoring, but is viewed as having significant potential. Some current problems associated with the method concern interference from $H_2O$ and $CO_2$ and high noise levels, which require the sample to be scanned several times to average the data. The papers of Plummer (Plummer, Logan, and Rollins 1990; Plummer 1991), Burrows (1990), and Stern and Thomson (1990) provide further information on source monitoring applications.

## SPECTROSCOPIC METHODS—ULTRAVIOLET MONITORING INSTRUMENTS

Analyzers operating in the ultraviolet–visible region of the spectrum have been used for pollutant and process gas monitoring for many years. The characteristics of light in the ultraviolet region of the spectrum (shorter wavelength, higher energy) lead to molecular electronic transitions when the light is absorbed. Because the energy associated with the absorption processes is greater than that in the infrared, it is easier to detect changes in absorption. Photomultiplier tubes traditionally have been used for this purpose; however, silicon-based linear photodiode array detectors are beginning to be applied in source monitoring instrumentation.

### Nondispersive Ultraviolet Photometers—Differential Absorption

When analyzing for gases in the ultraviolet region, a common approach is to utilize one or more of the narrow absorption bands in the UV spectrum (Figure 5-11). The instruments that are designed for the UV region operate in a manner somewhat different from the NDIR methods discussed previously. The technique of differential absorption is applied by measuring the difference in absorption between two wavelengths: a wavelength at which the molecule absorbs light and a wavelength at which it does not absorb light. The ultraviolet analyzer measures the degree of absorption at a wavelength in the absorption band of the molecule of interest (280 nm for $SO_2$ and 436 nm for $NO_2$, for example). This is similar to the NDIR method, but the major difference is that instead of a reference cell, a reference wavelength (in a region where $SO_2$ or $NO_x$ has minimal absorption) is utilized. The rationale behind this method again comes from the Beer–Lambert law.

In this nondispersive ultraviolet (NDUV) technique, a narrow band is chosen where the molecule does not have any spectral features (i.e., it does not absorb the light at that wavelength) and therefore $\alpha_{\text{reference}}(\lambda) = 0$. Then the measured value at this wavelength, $I_{\text{reference}}$,

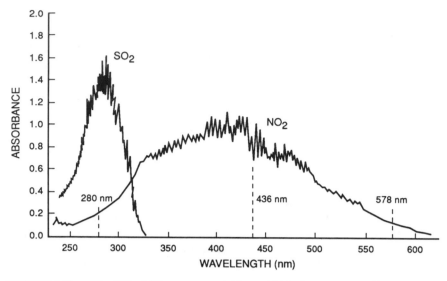

**FIGURE 5-11.**    The UV–visible spectrum of $SO_2$ and $NO_2$ (Saltzman, 1972).

gives the $I_0$ value as follows:

$$I_{\text{reference}} = I_0 e^{-\alpha(\lambda_{\text{reference}})c\ell} = I_0 e^0 = I_0 \tag{5-9}$$

The ratio between $I_{\text{sample}}$ and $I_0$ can then be taken to obtain a relation for the concentration:

$$\frac{I_{\text{sample}}}{I_{\text{reference}}} = \frac{I_0 e^{-\alpha c\ell}}{I_0} \tag{5-10}$$

$$c_{\text{sample}} = \frac{1}{\alpha\ell} \ln \frac{I_{\text{reference}}}{I_{\text{sample}}}$$

$$= k \log \frac{1}{\text{Tr}} \tag{5-11}$$

where $\ell$ is known (the length of the sample cell) and $k$ is a constant where $\alpha$ is either known or determined through appropriate calibration procedures. The value for the transmittance Tr is measured by the instrument.

This method of analysis is termed *differential absorption*, because measurements are performed at two different wavelengths. This method is not

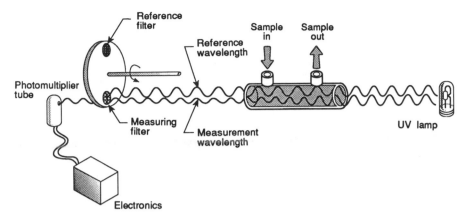

**FIGURE 5-12.**    Operation of a typical differential absorption nondispersive ultraviolet (NDUV) analyzer (opacity monitor).

limited to the ultraviolet but is also used in the infrared, particularly for $CO_2$ analyzers. The technique is also widely applied in other types of monitoring systems, including in-situ analyzers and remote sensors.

Figure 5-12 shows the operation of a typical differential absorption NDUV analyzer. Light from a mercury discharge lamp, hollow cathode, or other type of UV lamp passes through the sample cell to a set of band-pass filters rotating on a filter wheel and then to a photomultiplier tube. The measurement filter allows light centered in a narrow band at 285 nm to pass through to the detector. The other band-pass filter, the reference filter, allows light only in the region of 578 nm to pass through. Note (from Figure 5-11) the relative absorption of $SO_2$ in these regions. Light will be absorbed at 285 nm by the $SO_2$ in the sample gas, whereas no absorption will occur at 578 nm. The resultant photomultiplier signals are amplified and the logarithm of the reciprocal of the transmittance gives an output that is directly proportional to the $SO_2$ concentration, as shown in Equation (5-11).

Note that there is some interference by $NO_2$ at 578 nm, but there is also an equivalent $NO_2$ interference at the $SO_2$ absorption wavelength of 285 nm, so the net interference is negligible. Also, the concentration of $NO_2$ in a combustion gas is small (approximately 5% of the NO level) and the $NO_2$ absorption coefficient at 578 nm is small relative to that of $SO_2$ at 285 nm, which further reduces the effect of this interference.

Problems associated with lamp aging, power supply fluctuations affecting light intensity, and temperature effects on hollow cathode gas discharge tubes can result in short-term noise or long-term drift. These problems are minimized by modifying the basic system shown in Figure 5-12. Instrument manufacturers add a reference cell or split the UV beam as a continuous reference for the lamp output. Also, recent improvements in detectors and, in some cases, the use of solid-state detectors such as gallium arsenide phosphide (GaAsP) have improved overall instrument performance.

The differential absorption UV instruments have proven to be quite reliable in source monitoring applications. They are also used frequently by source testing firms in mobile vans because of their compact nature and relative freedom from interferences. As with all extractive system analyzers, particulate matter must be removed from the sample gas before the gas enters the instrument. It is not necessary, however, to remove water vapor in some systems designed to measure on a hot–wet basis. A heated sample line and heated cell prevent condensation in such systems. Because water vapor does not absorb light appreciably in this region of the ultraviolet spectrum, little interference occurs.

### Ultraviolet–Visible Spectrophotometers—Application of Linear Photodiode Array Detectors

With the advent of the photodiode array detector, a spectrometer capable of measuring multiple gases can now be constructed with almost the same simplicity as a photometer. This detector is composed of a series of silicon-based photodiodes that are sensitive to light. A diode array can have from 128 to over 4000 diode elements, each separated by about 25 $\mu$m. The array is first set up by charging each diode element with a reverse voltage to establish a barrier to electron flow. When light strikes the $n$-type silicon of the diode, photon-generated electrons discharge the diode. Each of the diodes is then recharged, the voltage necessary to recharge it being a measure of the light intensity on the element.

By using the diode array detector in conjunction with a diffraction grating, an instrument can be designed to obtain a spectrum in a few seconds (Figure 5-13). Using a holographic grating (a concave grating that separates the light beam into its component wavelengths), the light will fan out onto the diode array. The grating can be focused to spread out the light so that each diode element is sensing only a certain region of the spectrum. In essence, each diode is looking at only a few wavelengths, but the whole array will look at the total spectrum. Now, if sample gas is placed in the sample cell, light will be absorbed at different wavelengths and the diode elements that monitor those wavelengths will sense that reduction. So instead of having a spectrometer that disperses the light and then scans each wavelength with one detector, with a diode array, the light

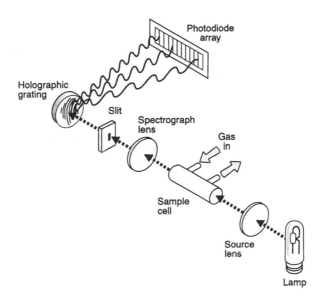

**FIGURE 5-13.**    A photodiode array spectrometer.

is dispersed and the absorption at each wavelength can be measured simultaneously with over 1000 diodes on the array.

The technique has great potential and can be utilized in both extractive analyzers (Saltzman 1990) and in-situ systems (Durham et al. 1990). The method obviously has the capability of measuring several gases at one time.

## SPECTROSCOPIC METHODS—LUMINESCENCE METHODS

Luminescence is a phenomenon caused by the emission of light from an excited molecule. There are three types of luminescent processes that are used in source monitoring analyzers:

1. Photoluminescence (fluorescence)
2. Chemiluminescence
3. Flame photometry.

Photoluminescence is the release of light after a molecule has been excited by light energy. Chemiluminescence is the emission of light from an excited molecule created in a chemical reaction. In flame photometry, the atoms of a molecule are excited to luminescence in a hydrogen flame. Analyzers utilizing the effects of luminescence can be very specific for a given pollutant and can have a greater sensitivity than some of the absorption or electrochemical methods.

### Fluorescence Analyzers for $SO_2$ Measurement

Fluorescence is a photoluminescent process in which light energy is absorbed at one wavelength and emitted at a different wavelength. In this process, the excited molecule will remain excited for $10^{-8}$ to $10^{-4}$ seconds. Upon the loss of energy by vibrational–rotational energy relaxation processes (internal conversion), the molecule goes to a lower energy state. It will then revert to its lowest energy state by emitting light, that is, it fluoresces. An energy-level diagram, given in Figure 5-14, illustrates the mechanism involved in fluorescence emission.

The molecule is in a high vibrational state after an electronic transition occurs (e.g., energy state $S_0 \rightarrow S_1$). Upon loss of some of the energy by vibrational relaxation, the molecule reverts to a lower energy state by emitting light (e.g., $S_1 \rightarrow S_0$) of wavelengths longer than those originally

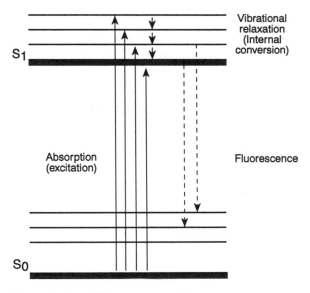

**FIGURE 5-14.**    Energy levels and fluorescence emission.

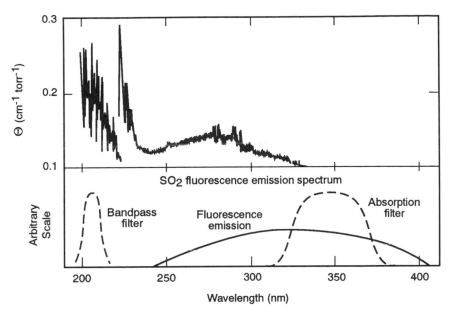

**FIGURE 5-15.**    Fluorescence in $SO_2$ (Strickler and Howell 1968; Okabe, Splitstone, and Ball 1973).

absorbed. The fluorescence process can be expressed as:

$$SO_2 + \underset{210\ nm}{h\nu} \rightarrow \underset{\substack{Excited \\ molecule}}{SO_2^*} \rightarrow SO_2 + \underset{240\text{–}410\ nm}{h\nu'}$$

where $SO_2^*$ symbolizes the excited molecule. The fluorescence spectrum for $SO_2$, shown in Figure 5-15, also illustrates this point.

To use this phenomenon to measure $SO_2$ concentrations, a sample is irradiated with light in the UV, in the range of 210 nm, and the emitted longer-wavelength fluorescent radiation is measured.

Commercially available $SO_2$ fluorescence analyzers have been developed to measure at ambient-level and source-level concentrations. A typical configuration is shown in Figure 5-16. Here, the light from either a continuous or pulsed ultraviolet light source is filtered to a narrow region, centered near 210 nm. The fluorescent radiation is measured at right angles to the sample, using a photomultiplier tube or other detector. A band-pass filter is then used to select a portion of the fluorescent radiation for measurement in the range of 310 to 370 nm, because interferences could occur over the wider range of the fluorescence emission spectrum.

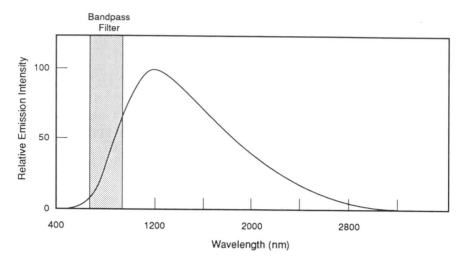

**FIGURE 5-17.**    The chemiluminescent emission spectrum of $NO_2{}^*$ (Clough and Thrush 1967).

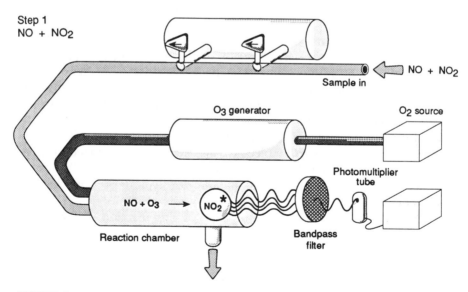

**FIGURE 5-18.**    Operation of a chemiluminescence analyzer: measurement of sample NO.

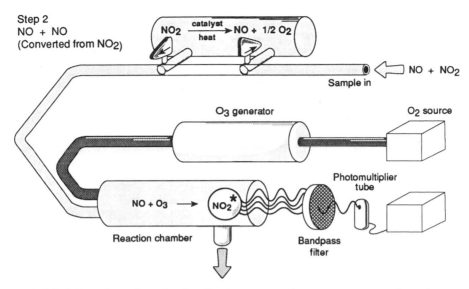

**FIGURE 5-19.** Operation of a chemiluminescence analyzer: measurement of sample NO + NO$_2$.

at lower temperatures. The NO produced is then reacted with the O$_3$ and the chemiluminescence is measured to give a total NO + NO$_2$ (NO$_x$) reading (Figure 5-19).

Step 1 of a typical measurement sequence is for the sample gas to bypass the converter and react in the reaction chamber.

Step 1:   [NO] + NO$_2$

In step 2 of the sequence, the sample gas is sent into the converter chamber and the NO$_2$ present is reduced to NO. So the sample sent to the reaction chamber in step 2 is as follows.

Step 2:   [NO] + [NO (converted from NO$_2$)] = [NO$_x$]

To obtain the amount of NO$_2$ in the sample, the results of step 1 are subtracted from those of step 2:

$$[NO_2] = (\text{Step 2}) - (\text{Step 1})$$

The sample gas can be switched automatically in and out of the converter to give sequential readings for both NO and NO$_2$, but this is usually an optional feature. For low levels of $NO_2$, as is usually experienced in combustion source applications, a negative value is sometimes obtained for the $NO_2$ concentration, particularly if the instrument is poorly calibrated.

Ammonia or other oxides of nitrogen can bias the $NO_x$ data. Ammonia will oxidize to NO in the converter and other nitrogen oxides will reduce to NO. Using a molybdenum converter, which operates at lower temperature, can minimize the problem. Also, because ammonia is very soluble in water, it will drop out with the moisture in a conditioning system.

The chemiluminescent radiation also can be quenched, as in fluorescence analyzers (Matthews 1977; Tidona 1988). The reaction chamber is commonly held under reduced pressure to lessen the effects of quenching. Also, quenching can be reduced by flowing $O_3$ into the sample chamber at a rate greater than the sample rate. The resulting dilution gives a relatively constant background gas composition. Through the incorporation of these techniques, the chemiluminescence method has been proven to be reliable and is often the method of choice for $NO_x$ measurements in source monitoring systems.

### Flame Photometric Analyzers for Sulfur Compounds

Another luminescence technique used to detect gaseous pollutants is that of flame photometry. Flame photometric analyzers have been used primarily in ambient air sampling, but have been applied to stationary source sampling by using sample dilution systems.

Flame photometry is a branch of spectrochemical analysis in which a sample is excited to luminescence by reactions occurring in a hydrogen flame. Instead of using an ultraviolet light source to excite the $SO_2$ molecule, as in fluorescence, the hydrogen flame is used to produce excited sulfur atoms (Farwell and Rasmussen 1976). These excited sulfur atoms will emit light in a band of wavelengths centered at about 394 nm, which is specific to sulfur. Compounds such as $H_2S$, $SO_3$, and mercaptans will contribute to the ultraviolet emission, giving a measure of the total sulfur content of the sample stream. Selective determination can be made of each compound by using scrubbers or chromatographic techniques.

A disadvantage of the flame photometric method is that hydrogen gas is necessary for the flame. Restrictions concerning hydrogen flames and hydrogen cylinders at some plants may preclude the use of this technique.

## ELECTROANALYTICAL METHODS

The instruments discussed in previous sections rely on spectroscopic, electro-optical techniques to monitor gas concentrations. Another class of instruments, based on electroanalytical methods of measurement, has found great utility in source monitoring applications. There are five distinct types of electroanalytical methods used in source monitoring:

1. polarography
2. potentiometry
3. electrocatalysis
4. amperometric analysis
5. conductivity.

Polarographic analyzers can be inexpensive and portable, ideal for inspection applications. Potentiometric analyzers using ion-selective electrodes have been successfully applied to measuring hydrochloric acid (HCl) and hydrofluoric acid (HF). The electrocatalytic, or high-temperature fuel-cell method as it is often called, is used to monitor oxygen only. Both extractive and in-situ monitors are available using this technique. The methods of amperometric analysis and conductivity are less widely used and are subject to a number of interferences.

### Polarographic Analyzers

Instruments have been developed that utilize the principles of polarography to monitor $SO_2$, $NO_x$, CO, $O_2$, and other gases. Polarographic analyzers (alternatively called voltammetric analyzers or electrochemical transducers) select from a range of electrodes and electrolytes to provide this versatility.

The transducer in these instruments is usually a self-contained electrochemical cell in which a chemical reaction with the pollutant molecule takes place. Two basic techniques are used in the transducer: (1) a selective semipermeable membrane allows the pollutant molecule to diffuse to an electrolytic solution; (2) the current change produced at an electrode by the oxidation or reduction of the gas at the electrode is measured. For example, in a system designed to measure $SO_2$, the oxidation reaction that takes place is

$$SO_2 + 2H_2O \rightarrow SO_4^{2-} + 4H^+ + 2e^-$$

$$E_{298}^0 = 0.17 \text{ V}$$

where $E_{298}^0 = 0.17$ V is the half-cell voltage potential.

Figure 5-20 illustrates the operation of a typical $SO_2$ electrochemical transducer. The oxidation–reduction reaction occurs at the sensing electrode because the counter-electrode material has a higher oxidation po-

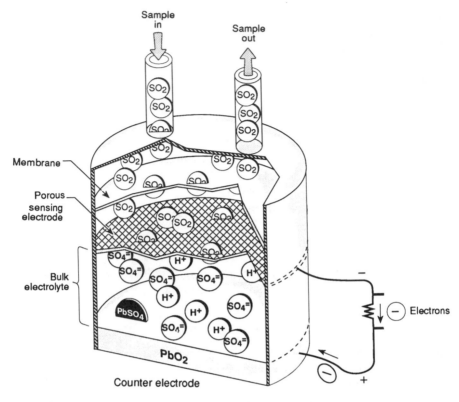

**FIGURE 5-20.** Operation of an electrochemical transducer designed to measure $SO_2$ concentrations.

tential than that of the species being reacted. In the cell, the sensing electrode has a potential equal to that of the counter electrode minus the voltage drop across the resistor. The sensing electrode is electrocatalytic in nature and, being at a high oxidation potential, causes the oxidation of the pollutant with a consequent release of electrons.

The reaction that takes place at the counter electrode is

$$PbO_2 + SO_4^{2-} + 4H^+ + 2e^- \rightarrow PbSO_4 + 2H_2O$$

$$E^0_{298} = 1.68 \text{ V}$$

where lead oxide ($PbO_2$) is converted to lead sulfate ($PbSO_4$).

The half-cell potential of 1.68 V is greater than $+0.17$ V for the oxidation of $SO_2$ to sulfate ($SO_4^{2-}$), so a net potential difference will develop. Similar oxidation–reduction reactions occur for different pollutants and electrode–electrolyte systems. Figure 5-17 shows that the operation of these systems involves (1) diffusion of the pollutant gas through the semipermeable membrane, (2) dissolving of the gas molecules in the thin liquid film, (3) diffusion of the gas through the thin liquid film to the sensing electrode, (4) oxidation–reduction at the electrodes, (5) transfer of the charge to the counter electrode, and (6) reaction at the counter electrode. The voltage drop across the resistor due to the current flow can be easily monitored and correlated to the pollutant concentration.

There are two reasons why this type of system may be termed polarographic or voltammetric. In typical polarographic analyzers used in chemical laboratories, the electric current in the system is related to the rate of diffusion of the reacting species to the sensing electrode. It turns out that if the rate at which the reactant reaches the sensing electrode is diffusion controlled, the current will be directly proportional to the concentration of the reactant. This is known as Fick's law of diffusion:

$$i = \frac{nFADc}{d} = kc \qquad (5\text{-}12)$$

where $i$ = current
$n$ = number of exchanged electrons per mole of pollutant
$A$ = exposed electrode surface area
$F$ = Faraday constant (96,500 coulombs)
$D$ = diffusion coefficient of the gas in the membrane and film
$c$ = concentration of the gas dissolved in the electrolyte layer
$d$ = thickness of the diffusion layer
$K$ = constant

Fick's law is similarly used in laboratory polarographic analyzers.

The other reason this type of system is termed polarographic is that a retarding potential can be maintained across the electrodes of the system to prevent the oxidation of those species that are not as easily oxidized. The difference between electrochemical transducers used for source monitoring and those used in the chemical laboratory is that in the laboratory instruments, an external potential is applied to the system until the decomposition potential of a given species is reached and an oxidation–reduction reaction occurs. By varying the potential, both qualitative and quantitative information can be obtained about the composition of a solution.

The polarographic technique is temperature dependent and any system, either permanent or portable, should be designed to maintain the cell at

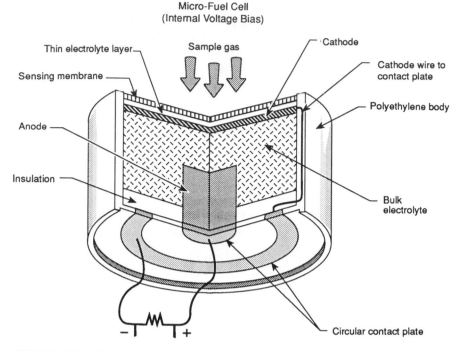

**FIGURE 5-21.**    Construction of an $O_2$ electrochemical cell.

constant temperature. It is also extremely important that the sample gas be properly conditioned before entering the analyzer. Particulate matter and condensed stack gases will easily foul the cell membranes, requiring the cell to be refurbished or replaced.

The cells themselves can come in a number of configurations, depending upon the manufacturer. These systems are small and portable and, compared to practically all other source monitoring instruments, they are the least expensive. These are two factors that make them ideal as inspection instruments or as warning monitors.

Electrochemical oxygen monitors used for detecting air inleakage in ducts and flues are quite popular. The configuration of a typical replaceable cartridge used for oxygen measurements is shown in Figure 5-21.

This technique is not as readily applied to permanent monitoring installations. As with a battery, the chemicals in the cell will eventually be consumed and the cell will then have to be either refurbished or replaced. For 24-hour operation in a permanent installation, replacement may be required every 3 to 6 months or less. This is a maintenance expense and requires careful attention by the instrument operator to note when degradation of the instrument performance begins to occur.

### Potentiometric Analyzers using Ion-Selective Electrodes

Potentiometric methods are used primarily to measure acid gases, where ion-selective electrodes can determine the change of an ion concentration in a buffered chemical solution. The technique requires more attention than the spectroscopic methods because of the recurring need to replace solutions and to maintain the analyzer plumbing. Although more popular in Germany and other European countries, this method has seen some application in the United States, particularly in ambient and source monitoring of HF at aluminum refineries.

In the potentiometric method, two electrodes are used to determine changes of the electromotive force of an electrolytic cell. One electrode serves as a reference electrode and the other serves as an ion-selective electrode, sensitive to the ion of interest (such as $Cl^-$ or $F^-$). The reference electrode maintains a potential that is independent of the solution composition.

Various stratagems are employed to bring the sample gas into contact with the buffered absorber solution. The absorber solution can be atomized or injected into specially designed mixing chambers and then run continuously or cyclically past the electrodes. The potential difference across the electrodes will vary with the ion concentration (activity). Sample gas and absorbing solution flow rates are critical, as are requirements for maintaining a constant sample temperature and pH balance. One advantage of the method is that an analyzer can be calibrated or audited by injecting a standard solution rather than a calibration gas.

### Electrocatalytic Analyzers for Measuring Oxygen

A method used for the determination of oxygen has been developed as an outgrowth of fuel-cell technology. The so-called fuel-cell oxygen analyzers are not actually fuel cells, but simple electrolytic concentration cells that use a special solid catalytic electrolyte to aid the flow of electrons. These analyzers are available in both extractive and in-situ configurations. This versatility of design has made them popular for monitoring diluent oxygen concentrations in combustion sources as part of permanent CEM systems.

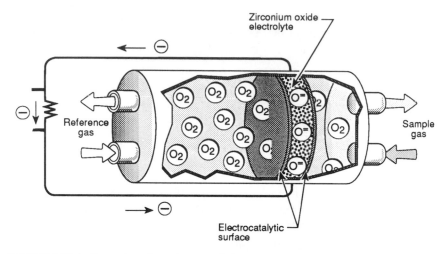

**FIGURE 5-22.**    Operation of an electrocatalytic oxygen analyzer.

In this method, a special ceramic material [zirconium oxide ($ZrO_2$)] coated with a thin layer of platinum] serves as an electrolyte to allow the transfer of oxygen from one side of a cell to another (Figure 5-22). In the cell, the oxygen concentration in the reference side is maintained at 21%. When sampling combustion gases, the oxygen concentration in the sample side will be less than that in the reference side (e.g., 3–6%).

When $ZrO_2$ is heated to 850° C, oxygen ions can migrate through the material. The thin film of platinum will catalyze the process, allowing oxygen to hop through the structure as $O^{2-}$ ions until they reach the other side. This hopping occurs because the zirconium ions form a relatively perfect crystal lattice in the material, whereas the $O^{2-}$ ions do not, resulting in vacancies in the structure. Heating the zirconium oxide allows the vacancies and $O^{2-}$ ions to move about. The oxygen ions migrate to the electrode on the sample side of the cell, release electrons to the electrode, and emerge as oxygen molecules.

This process continues until the concentration, or partial pressure, of oxygen becomes the same in both sides of the cell. In other words, the process continues until the chemical potentials become identical and the system comes into equilibrium. In practice, however, the sample side of the cell is continually flushed with new sample gas, not allowing the concentrations to become identical. As a result, a continual flow of electrons will occur across the resistor, as $O^{2-}$ ions continually move across the $ZrO_2$ in a vain attempt to establish equilibrium. The electromotive force (emf) of this process, expressed in terms of the oxygen partial pressures, is given as

$$\text{emf} = \frac{RT}{4F} \ln \frac{P_{\text{ref}}(O_2)}{P_{\text{sample}}(O_2)} \qquad (5\text{-}13)$$

where 
$R$ = ideal gas law constant
$F$ = Faraday's constant
$T$ = cell temperature
$P_{\text{ref}}(O_2)$ = partial pressure of $O_2$ in reference side of cell
$P_{\text{sample}}(O_2)$ = partial pressure of $O_2$ in sample side of cell

This emf can be measured. If the temperature is stabilized and the partial pressure of the oxygen on the reference side is known, the percentage of oxygen in the sample can be easily obtained. This phenomenon is used in some high-temperature fuel cells, where oxygen in the sample side is kept at a low concentration by reacting it with some fuel. In this way, an electric current is produced directly by combustion, not by using a mechanical electric generator. However, this points out one problem with the method in gas analysis. If CO, hydrocarbons, or other combustible materials burn at the operating temperature of the device, this results in a lowering of the oxygen concentration in the sample cell and gives readings lower than true. However, in most combustion sources, such gases are present at levels of parts per million and would provide a negligible error compared to the concentrations of oxygen measured on a percentage basis.

### Amperometric Analyzers

Amperometric analysis is a technique used in a few instruments developed for both ambient and source monitoring. These analyzers (also called coulometric analyzers) measure the number of coulombs required to produce a chemical reaction. Typically, amperometric analyzers measure the current in an electrochemical reaction, such as

$$SO_2 + 2H_2O + Br_2 \rightarrow H_2SO_4 + 2\,HBr$$

In this example, bromine is generated by an electrical current, which reacts with the $SO_2$ in the sample. The amount of current necessary to reach the reaction end point is proportional to the $SO_2$ concentration. However, amperometric instruments are susceptible to interferences from compounds other than those of interest. Problems with the necessary chemicals and associated plumbing also have made the application of these systems somewhat limited unless trained technicians are available to maintain them. The technique sees its greatest use for monitoring $SO_2$, $H_2S$, and mercaptans in Kraft pulp mill operations.

### Conductimetric Analyzers

Conductimetric analyzers sense the change in the electrical conductivity of a liquid reagent after the sample gas has been allowed to react with it. In continuous measurements, both the gas and the reagent are continuously mixed. The flow rates of both the gas and liquid streams must be kept constant for accurate measurements. Gases other than that being measured may also affect the conductivity of the reagent, so care must be given to the application of the method.

## PARAMAGNETIC TECHNIQUES
## FOR MEASURING OXYGEN

Molecules can be either attracted or slightly repelled by a magnetic field. When they are attracted, they are said to be paramagnetic in character. When they are repelled, they are termed diamagnetic. Most materials are diamagnetic when placed in a magnetic field, but a few materials are paramagnetic. Paramagnetism arises when a molecule contains unpaired electrons. Most materials will have paired electrons (one electron spinning in a clockwise direction for every one spinning in a counterclockwise direction) and therefore are diamagnetic. The oxygen molecule, however, is paramagnetic, having two unpaired electrons that spin in the same direction to give the molecule a permanent magnetic moment. When an oxygen molecule is placed near a magnetic field, the molecule is drawn to the field and the magnetic moments of the electrons become aligned with it. This striking phenomenon was first discovered by Faraday and forms the basis of the paramagnetic method for measuring oxygen concentrations.

The commercial paramagnetic analyzers are used only in conjunction with extractive systems. Water and particulate matter have to be removed before the sample enters the monitoring system. It should be noted that NO and $NO_2$ are also paramagnetic and may cause some interference in the instrument if high concentrations are present.

There are three measurement techniques developed that take advantage of this phenomenon:

1. thermomagnetic
2. magnetomechanical
3. magnetopneumatic techniques

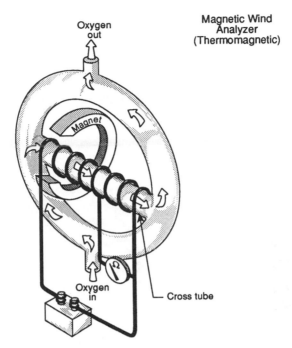

Oxygen
out

Magnetic Wind
Analyzer
(Thermomagnetic)

Magnet

Oxygen
in

Cross tube

**FIGURE 5-23.**    Operation of a thermomagnetic oxygen analyzer.

## Thermomagnetic Instruments

Analyzers employing the thermomagnetic method are frequently termed magnetic-wind instruments. They are based on the principle that the paramagnetic attraction of the oxygen molecule decreases as its temperature increases. A typical analyzer utilizes a cross-tube type of arrangement with a heated filament wire (Figure 5-23). A strong magnetic field covers one half of the coil. Oxygen contained in the sample gas will be attracted to the applied magnetic field and enter the cross tube. The oxygen then heats up and its paramagnetic susceptibility is reduced. The heated oxygen will then be pushed out by the colder gas just entering the cross tube. A flow of gas, or "wind," therefore continuously passes through the cross tube. This gas flow effectively cools the following section of the heated coil and changes its resistance. The change in resistance can then be detected in a typical Wheatstone bridge type of circuit and be related to the oxygen concentration.

Several problems can arise in the thermomagnetic method. The cross-tube filament temperature can be affected by changes in the thermal conductivity of the carrier gas. The gas composition should be relatively stable if consistent results are desired. Also, unburned hydrocarbons or other combustible materials may react on the heated filaments and change their resistance.

### Magnetodynamic Instruments

The magnetodynamic method is based on the effect that oxygen has in modifying the magnetic field in the vicinity of a permanent magnet. A specially constructed torsion balance, forming a "dumbbell" with diamagnetic glass spheres is suspended in a nonuniform magnetic field (Figure 5-24). The dumbbell spheres, being slightly diamagnetic, are pushed away from the most intense part of the magnetic field. When the gas surrounding the dumbbells contain oxygen, the spheres are pushed further out of the field due to the change in the field caused by the paramagnetic oxygen.

The system is first set up without sample gas so that light will reflect off a small mirror onto a photocell. In a feedback loop, a current is sent

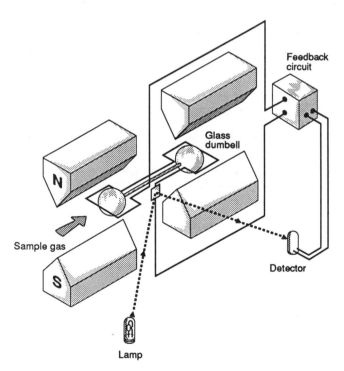

**FIGURE 5-24.**    Operation of a magnetodynamic oxygen analyzer.

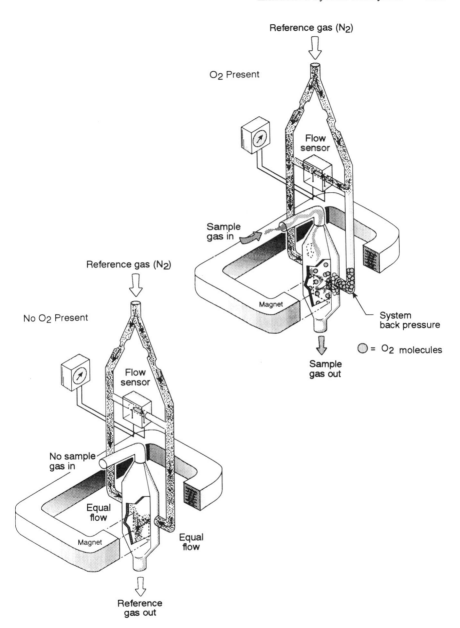

**FIGURE 5-25.**     Operation of a magnetopneumatic oxygen analyzer.

Marechal, A., Fortunato, G., and Laurent, D. 1985. A New Interferometric Optical Detector for Major Air Pollutants. In *Transactions—Specialty Conference on Continuous Emission Monitoring: Advances and Issues*. Air Pollution Control Association, Pittsburgh, pp. 251–260.

Matthews, R. D. 1977. Interferences in chemiluminescent measurement of NO and $NO_2$ emissions from combustion systems. *Environ. Sci. & Technol.* 11:1092–1094.

Okabe, H., Splitstone, P. L., and Ball, J. J. 1973. Ambient and source $SO_2$ detection based on a fluorescence method. *J. Air Pollut. Control Assoc.* 23:514–516.

Plummer, G., Logan, T. J., and Rollins, R. 1990. Fourier transform spectroscopy as a continuous monitoring method: A survey of applications and prospects. In *Proceedings—Specialty Conference on Continuous Emission Monitoring—Present and Future Applications*. Air and Waste Management Association, Pittsburgh, pp. 260–276.

Plummer, G. 1991. Measuring CAA Amendment hazardous air pollutants with Fourier transform infrared techniques. Paper presented at Air and Waste Management Association Meeting, Vancouver. Paper 91-58.7.

Podlenski, J., Peduto, E., McInnes, R., Abell, F., and Gronberg, S. 1984. *Feasibility Study for Adapting Present Combustion Source Continuous Monitoring Systems to Hazardous Waste Incinerators*, Vol. 1, *Adaptability Study and Guidelines Document*. EPA-600/8-84-011a.

Ryan, T. M. 1991. Electro-optic gas etalon absorption measuring instruments. Instrument Society of America, Paper 91-0408.

Saltzman, R. S., and Williamson, J. A., 1972. Monitoring stationary source emissions for air pollutants with photometric analyzer systems. In *Air Quality Instrumentation*, Vol. 1. Instrument Society of America, Pittsburgh, pp. 169–177.

Saltzman, R. 1990. A Process UV/Vis diode array analyzer for source monitoring. In *Proceedings—Specialty Conference on Continuous Emission Monitoring: Present and Future Applications*. Air and Waste Management Association, Pittsburgh, pp. 227–338.

Stern, L. P., and Thomson, J. W. 1990. Early field experience with an FTIR analyzer. In *Proceedings—Specialty Conference on Continuous Emission Monitoring: Present and Future Applications*. Air and Waste Management Association, Pittsburgh, pp. 277–291.

Strickler, S. J., and Howell, D. B. 1968. Luminescence and radiations transitions in sulfur dioxide gas. *J. Chem. Phys.* 49:1947–1951.

Strong, C. S. 1979. How the Fourier Infrared Spectrophotometer Works. *J. Chem. Education.* 56(10): 681–684.

Tidona, R. J. 1988. Reducing interference effects in the chemiluminescent measurement of nitric oxides from combustion systems. *J. Air & Waste Mgmt. Assoc.* 38:806–811.

Ueda, S., and Watanabe, A. 1978. The nondispersive infrared (NDIR) analyzer with improved selectivity and sensitivity for pollutants measurement. Paper presented at the Air Pollution Control Association Meeting, Houston. Paper 78-25.6.

## Bibliography

Allen, J. D., Billingsley, J., and Shaw, J. T. 1974. Evaluation of the measurement of oxides of nitrogen in combustion products by the chemiluminescence method. *J. Inst. Fuel* 12:275–280.

Beamish, C. J. 1981. A new analytic approach to accurate measurement of widely varying $SO_2$ emissions. In *Proceedings—Specialty Conference on Continuous Emission Monitoring: Design, Operation and Experience.* Air Pollution Control Association, Pittsburgh, pp. 165–175.

Beamish, C. J., and Kroepfl, D. 1985. TRS emission measurement using a microprocessor controlled gas chromatograph. In *Transactions—Specialty Conference on Continuous Emission Monitoring: Advances and Issues.* Air Pollution Control Association, Pittsburgh, pp. 103–113.

Bobeck, R. F. 1980. Electrochemical sensors for oxygen analysis. *Chem. Eng.* 7:113–117.

Dharmarajan, V., and Brouwers, H. J. 1987. Advances in continuous toxic gas analyzers for process and environmental applications. Paper presented at the Air Pollution Control Association Meeting, New York. Paper 87-62.8.

Dhyse, R. J., McGowan, G., and Cook, R. 1987. Application of FTIR to industrial gas measurements. In *Advances in Instrumentation.* Instrument Society of America, Anaheim.

Downey, J. E. 1987. On-line analyzers save time and money. *Instrum. Control Syst.* 4:31–37.

Federal Ministry for the Environment, Nature Conservation and Nuclear Safety. 1988. *Air Pollution Control Manual of Continuous Emission Monitoring.* Bonn, Germany.

Ginnity, B. 1988. Toxic gas monitor for accidental releases and process emissions. Paper presented at the Air Pollution Control Association Meeting, Dallas. Paper 88-46.1.

Herget, W. F., Jahnke, J. A., Burch, D. E., and Gryvnak, D. A. 1976. Infrared gas filter correlation instrument for in-situ measurement of gaseous pollutant concentrations. *Applied Optics* 15:1222–1228.

Herman, B. E., and Klein, F. 1990. Analysis of combustion stack nitric oxide emissions with non-dispersive ultraviolet detection. Paper presented at the Instrument Society of America Meeting. Paper 90-491.

Heyman, G. A., and Turner, G. S. 1976. Some considerations in determining oxides of nitrogen in stack gases by chemiluminescence analyzers. Paper presented at the Air Pollution Control Association Meeting, San Francisco. Paper 76-13.8.

Hodgeson, J. A., McClenny, W. A., and Hanst, P. L. 1973. Air pollution monitoring by advanced spectroscopic techniques. *Science* 182:248–258.

Jahnke, J. A., and Aldina, G. J. 1979. *Continuous Air Pollution Source Monitoring Systems—Handbook.* EPA 625/6-79-005.

Jones, C. 1986. Solid electrolyte gas sensors. Paper presented at the Instrument Society of America Meeting. Paper 86-2656.

Jones, D. G. 1985. Photodiode array detectors in UV–VIS spectroscopy: Part I. *Anal. Chem.* 57(9):1057A–1214A.

118     Continuous Emission Monitoring

Lang, C. J., Saltzman, R. S., and DeHaas, G. G. 1975. Monitoring volatile sulfur compounds in Kraft and sulfite mills. In *TAPPI 1975 Environmental Conference—Proceedings*. Technical Association of the Pulp and Paper Industry, Atlanta, 12-3-5.

Luft, K. F., Kesseler, G., and Zorner. 1967. Non-dispersive infrared gas analysis with the Unor analyzer. *Chem. Ing.-Tech*. 29(16):937–952.

Rieger, P. L. 1989. Application of FTIR spectrometry to motor vehicle exhaust measurements. Paper presented at the Air and Waste Management Association Meeting, Anaheim. Paper 89-34B.5.

Rollins, R., Logan, T. J., and Midgett, M. R. 1988. An evaluation of current instrumentation for continuous monitoring of hydrogen chloride emissions from waste incinerators. Paper presented at the Air Pollution Control Association Meeting, Dallas. Paper 88-137.1.

Saltzman, R. S., Small, J. R., and Steichen, J. C. 1989. New process UV/VIS diode array analyzer developments. Instrument Society of America, Paper 88-0107.

Shanklin, S., Rollins, R., Logan, T., and Midgett, R. 1989. HCl CEMS: Feasibility and reliability for municipal waste combustors. Paper presented at the Air Pollution Control Association Meeting, Anaheim. Paper 89-6.5.

Shanklin, S. A., Steinsberger, S. C., Logan, T. J., and Rollins, R. 1990. Evaluation of HCl measurement techniques at municipal and hazardous waste incinerators. In *Proceedings—Specialty Conference on Continuous Emission Monitoring: Present and Future Applications*. Air and Waste Management Association pp. 188–201.

Stevens, R. K., and Herget, W. F. 1974. *Analytical Methods Applied to Air Pollution Measurements*. Ann Arbor Science Publishers, Ann Arbor.

Vogelsang, R. F. 1987. Second Generation Analyzer Cross Flow modulation techniques—Precision, sensitive, continuous measurements. Paper presented at the Instrument Society of America Meeting. Paper 87-1091.

Yeh, J. T. Y. 1986. On line composition analyzers. *Chem. Eng*. 1:55-68.

# 6

# In-Situ Monitoring Systems for the Measurement of Gas Concentrations and Flue-Gas Velocity

Problems associated with extractive monitoring systems have led to the development of monitors that can directly measure source-level gas concentrations in the stack. These so-called in-situ systems do not modify the flue-gas composition and are designed to detect gas concentrations in the presence of particulate matter. In-situ units are constructed to withstand environmental conditions at stack locations and have fewer subsystems than do extractive systems.

In-situ systems originally were designed to measure combustion source flue gases having pollutant gas concentrations on the order of 500 ppm and greater. They have been proven to be successful and continue to operate well if maintained under appropriate quality assurance programs. However, with the advent of gaseous emission control systems and the resultant lowering of flue-gas pollutant concentrations, some in-situ analyzers have had difficulty in meeting certification requirements.

New interest is being shown in-situ systems due to advances in electro-optical design and the application of new calibration techniques that have overcome a number of earlier limitations. The application of newer technologies also promises to offer a wider range of in-situ analyzers in the near future. Several of these analytical techniques are discussed in Chapter 5, which can be referred to for additional information.

## TERMINOLOGY AND PRINCIPLES

In-situ monitoring systems are categorized as point monitoring systems or path monitoring systems. These are illustrated in Figures 6-1–6-3. Several

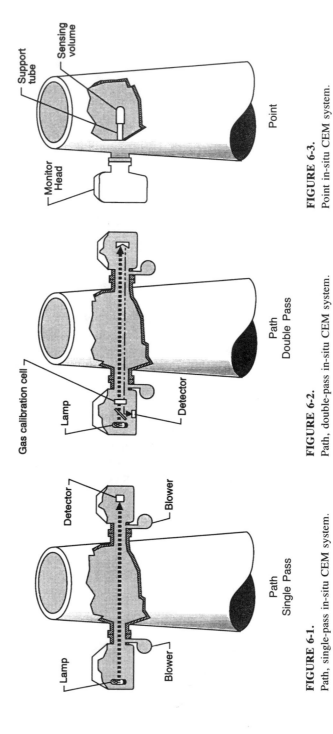

Types of In-Situ Analyzers

**FIGURE 6-1.**
Path, single-pass in-situ CEM system.

**FIGURE 6-2.**
Path, double-pass in-situ CEM system.

**FIGURE 6-3.**
Point in-situ CEM system.

other terminologies are also used. The point monitors have been referred to as *in-stack* monitors and the path monitors as *cross-stack* monitors.

Point in-situ systems (Figure 6-1) perform measurements at a single point in the stack, as do simple extractive system probes. The point, however, may extend from 5 to 10 cm or, in some electro-optical systems, to over 1 m when low gas concentrations are to be measured. The sampling path will still be relatively short compared to the stack or duct diameter, so consideration must be given to the length and placement of the probe if gas concentrations are stratified.

Path monitors (Figures 6-2 and 6-3) measure over a distance in a stack or duct that is usually equivalent to the internal stack diameter. In some cases, a pipe may be used for support or calibration purposes and may cut off part of the measurement path. Path monitors use electro-optical or acoustical techniques, where light or sound is transmitted through the flue gas. The effects of the flue gas on the transmission are used to provide a measurement of flue-gas parameters. There are two basic types of path system: single pass and double pass.

## Single-Pass Systems

Single-pass systems locate a transmitter and a detector on opposite ends of the duct or stack (Figure 6-2). Because light or sound is transmitted through the flue gas once only, these systems are termed single pass. [*Note:* in remote-sensing terminology, this arrangement is referred to as being *bistatic* (Vaughan 1991).]

Single-pass systems are used for the measurement of gas concentrations, flue-gas velocity, and opacity. The single-pass designs cannot be audited easily with certified cylinder gases (one of the audit options of the EPA Appendix F quality assurance requirements), a factor that has led to a decrease in their popularity.

## Double-Pass Systems

Double-pass systems locate both the transmitter and detector on one end of the duct or stack (Figure 6-3). To do this, a light beam is folded back on itself using a retroreflector. The light beam traverses the sample path twice, going from the transmitter-receiver ("transceiver") to the reflector and then back to the analyzer. (*Note:* in remote-sensing terminology, this arrangement is referred to as a *unistatic* configuration.)

Double-pass systems are used for measuring both gas concentrations and flue-gas opacity. The opacity monitors that meet EPA specifications are primarily double-pass systems (which are discussed in detail in Chapter 7).

## Optical Depth

When measuring over a path, a path-integrated concentration is obtained. The term "optical depth" is applied to such measurements, where optical depth is the product of the gas concentration and the measurement path, or

$$\text{optical depth} = c_s d_m \tag{6-1}$$

where $c_s$ is the gas concentration and $d_m$ is the distance that the light beam traverses through the flue gas. Optical depth is essentially an expression that compresses the real measurement path length into a 1-m length to give a normalized, equivalent concentration in units of ppm-meters (ppm-m).

Optical depth, expressed in units of ppm-meters, is useful in discussing the measurement capabilities of gas-monitoring path systems. For example, if a double-pass instrument has a range of 10,000 ppm-m, its full-scale reading (span) on a 5-m-diameter stack will be $10,000/(2 \times 5) = 1000$ ppm. Because optical depth is the product of both concentration and path length, one can obtain an equivalent optical depth by decreasing the path length and increasing the concentration.

Another example of the use of the optical depth expression is in determining appropriate gas concentrations for gas calibration cells. In a double-pass path monitor, for a 1 cm long calibration cell to find the $SO_2$ cell concentration necessary to give an optical depth equivalent to 10,000 ppm-m, consider the following calculation:

$$\text{optical depth} = 10,000 \text{ ppm-m}$$

$$c_{cell} d_{cell} = 10,000$$

$$c_{cell}(0.01)(2) = 10,000$$

$$c_{cell} = 500,000 \text{ ppm} \quad (50\% \text{ } SO_2)$$

In the case of small-diameter ducts or stacks, a path monitor may not be sufficiently sensitive to measure at low concentrations. For example, if a double-pass CO analyzer has a range of 5000 ppm-m and is installed on a 1-m-diameter duct, the minimum full-scale reading of the analyzer would be $5000/2 = 2500$ ppm. This range may not be appropriate if flue-gas CO concentrations were normally on the order of 100–200 ppm. There are, however, several solutions to this problem. The measurement path length

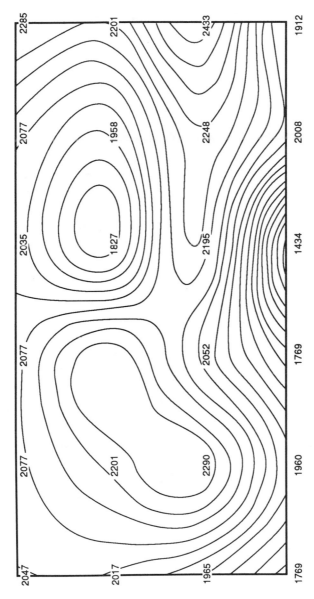

**FIGURE 6-4.** Gas stratification in a duct (Brooks and Williams 1976).

can be increased (or possibly reduced) by installing the path system in an alternate configuration, such as parallel to the stack or duct, or at an angle across the stack. In some installations, the gas can be bypassed through an external measurement tube, with a longer path length, to make the measurement.

Installing the path system at an angle can increase the path length to gain more sensitivity in the instrument. However, alignment may be more difficult, and the possibility exists that moving parts, bearings, and so on may wear unevenly when not placed in a horizontal position. Installing a system in a parallel configuration is not possible for most in-situ analyzers, because the transmitter and receiver must be external to the stack. It is possible, however, in systems that incorporate fiber-optic cables.

A bypass installation can solve many problems associated with in-situ stack measurements. Such an installation essentially converts an in-situ system into an extractive system. Here, the length of a bypass duct can be adjusted to be either shorter or longer than the stack diameter. An optical depth can be tailored to the measuring range of the instrument. Also, if water droplets are present in stack gas, the duct can be heated to minimize this interference with the measurement.

In cases where the gas concentrations are stratified (Figure 6-4) (Brooks and Williams 1976), it is frequently assumed that a path-integrated concentration will give a more representative value for determining stack emissions than a value obtained by a point monitoring system. This assumption may or may not be true, depending on the procedures used to define a representative sample. Both U.S. and international methods (U.S. EPA Reference Method and ISO Standard—ISO 9096) specify that reference-method measurements are to be taken at a minimum number of points located at centers of equal areas of the cross section.

In the case of Figure 6-4, a horizontal line average through the center of the duct differs from the area average by approximately 4%. In the case of S-shaped stratification or cyclonic flow, values may differ even more widely. Although, in many cases of stratified flow, a path-integrated concentration will be more representative than a point-determined concentration, the stratification patterns should still be characterized. These characterizations must consider stratification changes as a function of time, space, and process load.

## PATH IN-SITU ANALYZERS

Two principal techniques are applied in commercially available path analyzers. These are differential absorption spectroscopy and gas filter correlation spectroscopy. These techniques are also used in extractive system analyzers; however, different instrument configurations are necessary to apply these methods for in-situ measurement. As a result a number of different approaches to analyzer design are taken by instrument manufacturers.

### Differential Absorption Techniques

Differential absorption techniques are easily adapted to in-situ measurement. However, the measurement path length and the light-absorbing characteristics of the many gaseous species present must be considered in the selection of measurement wavelength. In spectral regions where absorption is high, too much light may be absorbed over a long stack path length ($> 10$ m in some cases, which could be doubled to 20 m if a double-pass system were used). Some light energy does need to reach the instrument detector, so a spectral band having a weak absorption coefficient may be more suitable in such situations.

Also, for path in-situ analyzers, particulate matter and water droplets in the flue gas will scatter the light, reducing the amount of light energy that can pass through the optical path to the detector. This should not be a problem for opacities less than 20–30% if the instrument is properly designed. For opacities lower than 30%, opacity fluctuations should not affect the instrument measurements when using the differential absorption technique. If the intensity of the light energy at the measuring wavelength is the same as that at the reference wavelength, then each intensity would be reduced by a constant factor and the reduction would cancel out as shown:

$$I = KI_{wp}$$

$$I_0 = KI_{0\,wp} \tag{6-2}$$

where $\quad I$ = measured light intensity at the measuring wavelength

$\quad I_0$ = measured light intensity at the reference wavelength

$\quad K$ = fraction of light attenuated by particulate matter in the gas stream

$\quad I_{wp}$ = hypothetical light intensity at the measuring wavelength without particulate matter present

$\quad I_{0\,wp}$ = hypothetical light intensity at the reference wavelength without particulate matter present

and taking the signal ratio ($I/I_0$), we have

$$\frac{I}{I_0} = \frac{K \times I_{wp}}{K \times I_{0\,wp}} = \frac{I_{wp}}{I_{0\,wp}} \tag{6-3}$$

This satisfies the requirement demanded of all in-situ analyzers: that particulate matter should not interfere in the analytical method. Interference caused by broadband absorption of water vapor or other molecular species should cancel similarly if the absorption of the interferant at the measuring and reference wavelengths does not differ significantly.

### Use of Band-Pass Filters

Double-pass path in-situ analyzers have been developed that use band-pass filters exclusively to obtain the measurement and reference wavelengths. Figure 6-5 illustrates the design used in one instrument, the Erwin Sick GM-21 $SO_2$ in-situ analyzer.

Three filters, having band-pass wavelengths centered at 313, 436, and 546 nm, are used. Referring back to Figure 5-11, note that there is some overlap of the $SO_2$ and $NO_2$ spectrum in the UV region. The instrument makes a measurement of light intensity at 546 nm that is reduced by $NO_2$ absorption and particle scattering, an intensity at 436 nm that is also reduced by the effects of $NO_2$ and particles, and an intensity at 313 nm

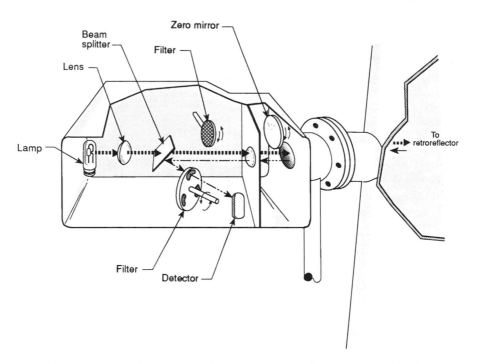

**FIGURE 6-5.**    A path double-pass in-situ analyzer used for the measurement of $SO_2$, $NO_2$, and opacity.

that is reduced by $SO_2$ absorption in addition to the effects of $NO_2$ absorption and particle scattering. A microprocessor takes these measured values and solves three simultaneous linear equations to obtain the pollutant parameters.

The unit consists of a filter wheel that continuously rotates the three filters, a low-pressure mercury vapor lamp, and a photodetector. A corner-cube retroreflector is located on the opposite side of the stack or duct, to reflect the beam transmitted from the lamp back to the detector. A mirror, external to the front lens of the transceiver assembly, is swung into the light beam every 15 min to simulate a measurement over a clean stack. Reference values obtained from the mirror in this position are used to compensate for changes in ambient temperature, aging of the light source, and aging of the photodetector. This mirror, in conjunction with an internal mirror, is also used to compensate for dirt accumulation on the transceiver window. Blowers on both the transceiver and retroreflector sides are used to keep the optical surfaces that are exposed to the flue gas clean. They also assist in cooling the units.

The Erwin Sick instrument basically serves as an $SO_2$ monitor because the $NO_2$ and opacity measurements are made more with the intent of correcting the $SO_2$ value than for supplying regulatory data. In the present design, light attenuation caused by particulate matter is not recorded as opacity, but as "extinction," a measure of the amount of light scattering per unit path length.

### Use of Diffraction Gratings
In the 1970s, the most widely marketed path in-situ system was a single-pass differential absorption unit that incorporated a diffraction grating to select appropriate wavelengths for the measurement of $SO_2$ and NO, and narrow band-pass filters for the measurement of $CO_2$ and CO. The characteristics and performance of this system are well documented in the literature (Bambeck and Huettemeyer 1974; Jahnke and Aldina 1979; Lord 1976, 1982; Nation 1981; Schuck 1981). Although a number of units continue in operation, the analyzer is no longer marketed. Basically, the difficulties associated with the calibration of a single-pass system created too much uncertainty in the measurements.

Recently, a more sophisticated differential absorption technique has been developed to perform either ambient or source pollutant measurements (Karlsson 1990; Biermann and Winer 1990). The technique is one of the "differential optical absorption spectroscopy" (DOAS) methods that are currently undergoing significant development. The difference between these and earlier differential absorption analyzers, is that in DOAS systems a relatively large portion of an absorption spectrum is analyzed, not just a few wavelengths. The technique is a dispersive method, not a nondispersive (NDUV) one that looks at only narrow spectral bands.

In a typical DOAS system, a diffraction grating is used to disperse the signal beam into its component wavelengths. A moving slit assembly is then used to scan the spectrum, as shown in Figure 6-6. The light intensity associated with each wavelength is measured by the detector as the slit moves over the diffracted spectrum. This information can then be processed and stored in a computer, as are the spectral data and calibration data. Concentration values are obtained by matching the two sets of information.

This technique has been commercially applied by ABB-OPSIS, a company based in Sweden. The instrument is a single-pass unit and can be configured to measure either ambient- or source-level pollutant concentrations. It can be used to monitor $SO_2$, NO, $NO_2$, $NH_3$, Hg, $H_2O$, $CO_2$, as well as other gases.

In the system, a high-pressure xenon lamp is used to send light over a fiber-optic cable to a stack-mounted transmitter. The light is projected through the flue gas and is detected by a transmitter that sends it back through a fiber-optic cable to the spectrometer. In the spectrometer, the

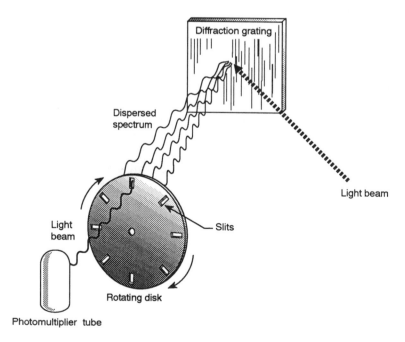

**FIGURE 6-6.**    Operation of a moving slit assembly in a differential absorption spectrometer.

light is diffracted from a grating and the spectrum is scanned. A computer collects the spectral data, compares it with stored reference spectra, and calculates concentrations for specified gases. A calibration can be performed by using external calibration cells and another set of fiber-optic cables to transmit and receive light through the cell.

### *Use of Diode Array Detectors*

A path double-pass in-situ instrument has been developed that uses a diode array detector instead of a grating to distinguish between different light wavelengths. This technique is also a differential absorption method and is described in Chapter 5, in conjunction with its application to extractive system analyzers. Erwin Sick, in its Model GM30, has adapted the method for in-situ measurements as shown in Figure 6-7.

In contrast to the differential absorption filter analyzer discussed previously, a photodiode array is used to measure $SO_2$ and NO. Opacity is measured using a beam splitter and separate photodetector. Using a deuterium lamp, ultraviolet and visible light is transmitted across the stack to a beam splitter. Part of the beam goes to the opacity photodetector and the other to a diffraction grating and the silicon-based photodiode array.

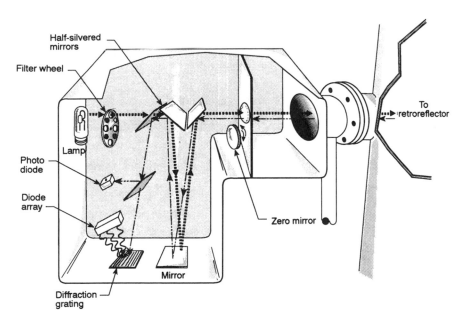

**FIGURE 6-7.**    Application of a photodiode array detector to measure $SO_2$, NO, and opacity in an in-situ CEM system.

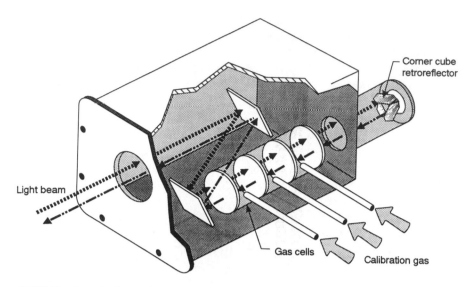

**FIGURE 6-8.**    Audit attachment for performance checks of a path in-situ monitor.

Light attenuated at the 550-nm wavelength is used to determine the flue-gas opacity. Light absorption in the region from 218 to 233 nm is used for measurement of $SO_2$ and NO. The analyzer uses an NO gas cell mounted on a rotating filter wheel to obtain spectral reference measurements.

An audit attachment is available for this instrument, which can be used to check the linearity of the gas measurements (Figure 6-8). The attachment can be bolted to the transceiver assembly by swinging the transceiver assembly on the air purge assembly flange, away from the port. The audit attachment contains a retroreflector that simulates the mirror on the opposite side of the stack. Three gas cells are positioned inside so that the light beam passes through them. Because each cell can be filled independently, only one calibration gas or audit gas is needed to make three different measurements. By sequentially filling each cell, three different optical densities are obtained for an audit or linearity check. To check the opacity measurement, neutral density filters can be used to attenuate the light beam.

### The Gas Filter Correlation Technique

The technique of gas filter correlation (GFC) spectroscopy was first applied for source emission measurements in in-situ monitoring systems

(Herget et al. 1976). The technique has undergone many developments since and is now successfully applied as well in extractive systems, as discussed in Chapter 5. In-situ GFC optical configurations have been designed that employ the technique either in a single-pass or a double-pass mode. Single-pass GFC systems continue to be marketed and do operate successfully, although they continue to suffer from the lack of suitable cylinder gas calibration techniques. The double-pass GFC systems do, however, show more promise.

A double-pass GFC monitoring system has been recently developed and is now commercially available. This system, developed and marketed by AIM, Inc., can measure a number of gases and uses both gas filter cells (applying the GFC technique) and band-pass filters (applying the differential absorption technique). The cells and band-pass filters are set into two rotating turret assemblies that step through a prescribed sequence of reference and stack measurements (Figure 6-9).

This instrument has had applications in monitoring for $SO_2$, NO, $CO_2$, CO, HCl, $H_2O$, HF, hydrocarbons, and opacity, being able to measure up to six different gases and opacity with one analyzer. The gas filter correla-

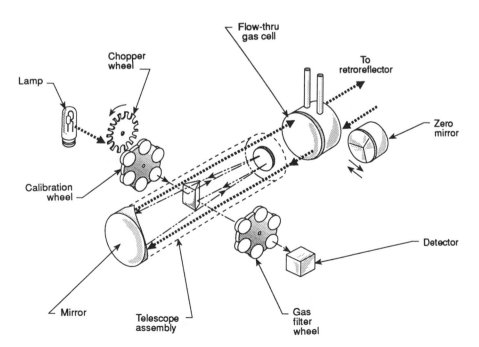

**FIGURE 6-9.**    A double-pass gas filter correlation in-situ system.

tion technique is used for the measurement of NO, CO, and HCl. Optical filters are used for gases such as $H_2O$, $SO_2$, $NO_2$, hydrocarbons, and $CO_2$.

A neutral reference cell, containing a nonabsorbing gas (such as $N_2$) is used in a manner similar to that discussed in the design of the extractive GFC analyzer (Chapter 5). As in the extractive GFC analyzer previously discussed, concentration measurements are obtained by comparing the absorption between the GFC and reference cells. To incorporate a mechanism for a daily calibration check, a mirror is located inside the transceiver assembly to provide an internal zero value. Internal, sealed, calibration cells inserted on the turrets can be used to give upscale readings for a daily calibration check. As an option, an internal flow-through gas cell can be used for conducting daily upscale calibration checks, or cylinder gas audits.

As another option, an audit gas cell (similar to that shown in Figure 6-8) can be attached to the transceiver assembly to perform a cylinder gas audit (Figure 6-10). This longer flow-through audit cell is 3 ft long and is bolted to the front of the transceiver. After the cell is attached, the light now reflects from a mirror at the end of this "simulated stack" to give measurements for reference gases, the optical depth being suitably corrected for the new path length. With proper conditioning, source gas could also be passed through the audit cell, to convert the in-situ monitor into an extractive analyzer.

A simple single-pass GFC monitor used for the measurement of CO is currently marketed by Land Combustion, Ltd. This analyzer consists of an

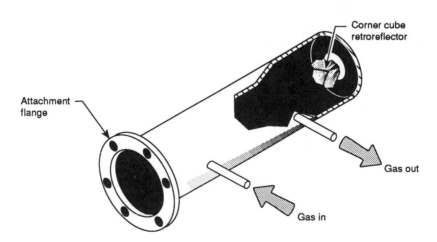

**FIGURE 6-10.**    Audit cell for an in-situ path monitor.

infrared transmitter and a receiver assembly. The receiver incorporates a detector that monitors the transmitted beam and a reference light source. To obtain a gas filter measurement, a mirror moves into place to block off the light beam transmitted from across the stack and reflects the light from the reference source to the detector, through a small gas filter cell that contains CO. This instrument provides no means for internal calibration and is best checked using an independent instrument.

## Advantages and Limitations of Path In-Situ CEM Systems

Path in-situ systems offer a number of advantages over extractive systems, but they also have some limitations. In practice, the double-pass path systems offer more flexibility in calibration than do the single-pass designs. For meeting regulatory requirements, the double-pass systems are generally preferred.

One of the principal marketing features of path in-situ analyzers is that a single instrument can monitor a number of gases. However, the cost of such a system can be comparable to the purchase price of three or four separate instruments combined in an extractive system. Daily operating costs, in principle, can be less than extractive systems costs, because zero and span gases may not be necessary for the 24-h checks (depending upon the applicable regulation). There are also fewer subsystems associated with in-situ monitors, so problems occurring with chillers, heat-traced lines, valves, and pumps are avoided.

Path monitors, however, do have a number of limitations. A path in-situ system can monitor only one flue or stack at a time; thus, time-sharing arrangements between different ducts or stacks are not possible, as they are with extractive systems. On the other hand, time-shared systems have their own limitations, so this problem may not be particularly significant.

More important, in-situ systems can be sensitive to environmental factors such as vibration and elevated temperatures. Vibration can loosen circuit boards and components and can misalign optical systems. High ambient temperatures may affect electronic circuit boards, necessitating a cooling or air conditioning system. High internal stack temperatures [greater than 260°C (500°F)] may result in radiative effects that interfere with infrared measurements. As mentioned earlier, high flue-gas opacities (> 30%) due either to particles or water droplets may reduce the beam intensity to such a degree that too little light will reach the detector for analysis. Lightning strikes can also be a problem.

Performing cylinder gas audits has been one of the principal problems associated with path in-situ analyzers. The limitation is most significant in the single-pass units, but can be resolved in the double-pass systems, as has been illustrated. Flow-through gas cell calibration techniques are similar in principle to calibration techniques used in transmissometers, which provide a systems check of the electronic and optical components of the transceiver. Therefore, the technique should be acceptable to most regulatory agencies.

## POINT IN-SITU ANALYZERS

Point in-situ analyzers offer a means of measurement that is, in some respects, a compromise between the path monitoring technique and extractive systems. In a point in-situ system, the measurement is made at a point, or a path that is short relative to the diameter of the stack or duct. In the case of stratified flow, this again brings up the problem of finding a representative point to monitor. It is more difficult to obtain different lengths for point monitor probes, and there is always a practical limit for their length. Nevertheless, point monitors have had many successful applications, particularly for measuring $SO_2$ and $O_2$ in combustion sources.

Both spectroscopic and electroanalytical techniques have been applied in point monitors. Differential absorption IR analyzers and second-derivative UV spectrometers are the most common electro-optical systems. The electrocatalytic zirconium oxide sensor used to measure $O_2$ is the most widely applied electroanalytical technique, although point electrochemical analyzers have also been developed for monitoring $SO_2$ and NO.

### Differential Absorption

Simple, point in-situ differential absorption instruments can be constructed by attaching a retroreflector to the end of a probe, rather than placing it on the other side of the stack. The retroreflector can be protected by a ceramic filter, which allows flue gas to diffuse through it, but stops particulate matter from entering the measuring cavity. As in extractive system probes, a deflection plate may be attached to minimize the direct impingement of particles onto the filter. The length of the measurement cavity will depend on the range specified for the instrument.

In such a system, light of the appropriate wavelengths is projected down the probe through a protective window, into the actual measurement cavity, which may be only a few centimeters in length. Light energy is consequently absorbed by the flue gas that has diffused through the filter into the cavity, and the remaining light is returned by the retroreflector to the detector in the transceiver assembly. A calibration port located at the connecting flange, attaches to a tube that enters the measurement cavity. Zero gas or calibration gas can be sent into the cavity to purge the flue gas when conducting a daily calibration check or a cylinder gas audit.

By selecting appropriate optical filters, differential absorption point monitors have been developed by Lear Siegler Measurement Controls for monitoring CO, $CO_2$, and $H_2O$ (Dec 1986).

### Second-Derivative Spectroscopy

The technique of second-derivative spectroscopy (Hager and Anderson 1970; Hager 1973; Williams and Palm 1974) has been used by analytical chemists to enhance absorption signals when faced with measuring low gas concentrations over short path lengths. Lear Siegler Measurement Controls has adapted the method in a point in-situ analyzer, where the gas is measured over a relatively short path length of 5–10 cm. Second-derivative spectroscopy is a unique method that involves scanning a spectral absorption peak and obtaining its second derivative with respect to wavelength at the peak maxima.

If the Beer–Lambert expression characterizes the absorption curve, it can be shown that its second derivative is proportional to the gas concentration (Hager and Anderson 1970; Hager 1973). For the Beer–Lambert expression given in Chapter 4,

$$I = I_0 e^{-\alpha c \ell} \tag{6-4}$$

where $I_0$ = the intensity of the light entering the probe
$I$ = the intensity of the light leaving the probe
$\alpha$ = wavelength-dependent molecular absorption coefficient
$c$ = concentration of the measured gas
$\ell$ = light path length through the gas

a second derivative with respect to wavelength of an expansion of the expression can be given as

$$\frac{d^2 I}{d\lambda^2} = -c\ell \frac{d^2\alpha}{d\lambda^2} I \tag{6-5}$$

Note that the second derivative of the intensity is proportional to the concentration $c$.

The second-derivative spectrometer operates by producing a signal that is proportional to

$$\frac{d^2 I}{d\lambda^2} \quad \text{at } \lambda_0 \qquad (6\text{-}6)$$

where $\lambda_0$ = the center of the absorption curve (Jahnke and Aldina 1979)

If this signal, $S$, is proportional to the second derivative, then it is also proportional to the concentration of the gas. Essentially then, the instrument gives a signal that is proportional to the gas concentration, which is of course what is necessary to make a monitoring instrument.

The signal produced by the instrument is given by

$$S = \frac{\delta}{4} \frac{d^2 I}{d\lambda^2} \qquad (6\text{-}7)$$

where $\delta$ = the scanning distance (for example, 1.8 nm in the Lear Siegler monitor)

Substituting the second derivative expression for the Beer–Lambert law, the resultant expression for the signal is

$$S = \frac{\delta^2}{4} \left[ -c\ell \frac{d^2 \alpha}{d\lambda^2} I \right] \qquad (6\text{-}8)$$

or

$$\frac{S}{I} = Kc\ell \qquad (6\text{-}9)$$

where $K$ is a constant. This is the actual instrument output, which is proportional to both the gas concentration and the optical path length. By dividing $S$ by $I$, problems caused by variations in the source intensity, broadband absorption from other gases, or scattering by particulate matter are avoided. This results because a change in $I$ by a constant factor will induce an identical change in $S$. Determining the ratio of the two cancels out the effect.

The question then arises, how does the spectrometer produce a signal that is proportional to the second derivative of $I$, the measured light intensity? Figure 6-11 illustrates the basic optical system. In this system,

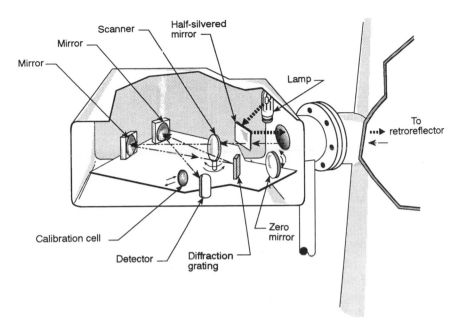

**FIGURE 6-11.**    Operation of a second-derivative point in-situ monitor.

light is sent from a UV lamp down the probe to the measurement cavity and is returned using a corner-cube retroreflector. In the transceiver, a diffraction grating selects the specific absorption wavelengths, but instead of just sitting on a specific wavelength as is done in differential absorption techniques, a wobbler, or oscillating lens, moves the returning beam back and forth across the diffraction grating to scan the absorption peak.

For $SO_2$ analysis, light is monitored in the region of 218.5 nm, which corresponds to the maximum of an $SO_2$ absorption peak in the ultraviolet. The wobbler scans the light over wavelengths from 217.8 to 219.2 nm, across the width of the absorption peak. This creates an oscillating signal that has a frequency twice that of the scanning frequency $f$.

Electronically, the analyzer tunes in on the frequency that is double that of the frequency of movement of the scanner, much like tuning in a radio. A station with a strong transmitter will produce a louder signal than a weaker station. A strong signal at this frequency indicates strong absorption and a high concentration of $SO_2$; a weak signal indicates a low concentration. The amplitude $S$ of the detector signal at the frequency $2f$ is proportional to $d^2I/d\lambda^2$ evaluated at $\lambda_0$ [Equation (6-6)] and therefore to the concentration [Equation (6-8)].

The performance of this instrument can be checked in three different ways, using either an internal gas cell, an external gas cell, or calibration gases.

1. *Internal gas cell.* A gas cell inside the instrument can be placed into the path of the light beam after a zero mirror moves into place (Figure 6-11). The cell contains both $SO_2$ and NO and can give an upscale reading to check the electro-optical system of the transceiver. The internal cell is often used to perform the daily calibration check of the instrument.
2. *External gas cell.* Independent gas calibration cells can also be inserted into the light path from the top of the instrument. These cells can also provide a performance check; however, their values are generally not certified and therefore have limited use for auditing.
3. *Calibration gases.* The best way of calibrating or auditing the instrument is to inject calibration gas into the measuring cavity (Figure 6-12). A port and internal sample line are provided for this purpose. The sample line is coiled inside the probe to allow the calibration gas to heat up to stack temperature before entering the cavity. A gas pressure sufficient to purge the stack gas is applied and is increased until a stable value is obtained.

In some situations, the ceramic filter may become "blinded" with particulate matter or coated with precipitate from a malfunctioning scrubber. The instrument may read the zero and calibration gas values correctly if sufficient pressure is applied. However, if it takes an extended period of time for the system to come back to the stack value, the problem should become evident and the filter should then be replaced or cleaned.

### Electrochemical Analysis

Techniques other than differential absorption and second-derivative spectroscopy have also been developed for the in-situ point measurement of $SO_2$. At present, two commercial systems using electrochemical techniques are available. Rosemount, Inc., markets an analyzer that uses a solid electrolyte to measure $SO_2$. This system is used in conjunction with an in-situ zirconium oxide probe. Land Combustion, Ltd., has available in-situ $SO_2/NO$ and NO point monitors that use liquid electrolytes.

#### A Solid Electrolytic Cell for Measuring $SO_2$
The Rosemount in-situ monitor sensing cell consists of a potassium sulfate solid electrolyte with platinum electrodes (Nelson 1985; Jones 1985a, b).

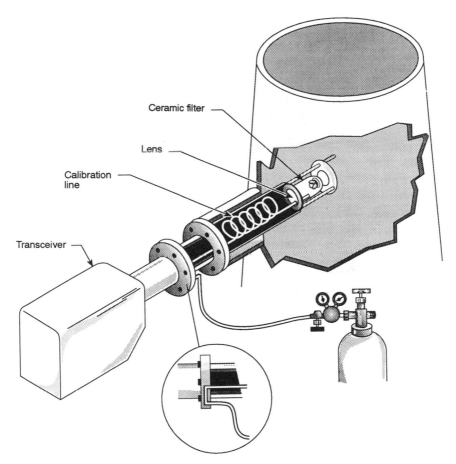

Ceramic filter

Lens

Calibration line

Transceiver

**FIGURE 6-12.**    Performance check of an in-situ point monitor using a calibration gas.

In contrast to the oxygen analyzer using the zirconium oxide electrolyte, where the electromotive force (emf) is generated by differing oxygen concentrations on each side of the electrolyte, the emf in the $SO_2$ analyzer is generated by differing potassium ion activities. This is done by using two potassium sulfate half-cells in the form of pellets positioned at the end of a probe, as shown in Figure 6-13.

The measuring half-cell is exposed to the flue gas, whereas the reference half-cell is exposed to a flow of reference gas. The mullite (aluminum silicate) barrier between the two cells is impervious to gas, but will allow the conduction of potassium ions. The voltage developed across the

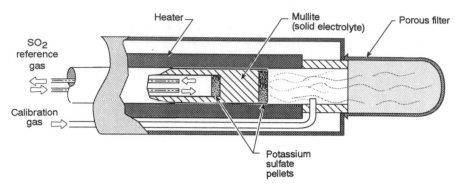

**FIGURE 6-13.**    Potassium sulfate ($K_2SO_4$) electrochemical point in-situ monitor used for measuring $SO_2$: probe cell design.

electrodes results from the flow of the potassium ions through the system. The potassium ion activity at each electrode is a function of both the $SO_2$ and $O_2$ concentrations. The Nernst relationship for this particular system is

$$\text{emf} = \frac{RT}{2F} \ln \frac{P''_{SO_2}}{P'_{SO_2}} + \frac{P''_{O_2}}{P'_{O_2}} + C \tag{6-10}$$

where    $F$ = Faraday's constant

$R$ = gas constant

$T$ = absolute temperature

$P''_{SO_2}$ = partial pressure of $SO_2$ at the reference electrode

$P'_{SO_2}$ = partial pressure of $SO_2$ at the sensing electrode

$P''_{O_2}$ = partial pressure of $O_2$ at the reference electrode

$P'_{O_2}$ = partial pressure of $O_2$ at the sensing electrode

$C$ = cell constant

In the analyzer, the partial pressure of the oxygen and the $SO_2$ are kept constant at the reference electrode, using reference gas, as is the cell temperature. By using a zirconium oxide in-situ analyzer to determine the in-stack $O_2$ concentration $P'_{O_2}$, the in-stack $SO_2$ concentration $P'_{SO_2}$ can be determined.

The system is calibrated with cylinder gas by sending gas down the probe to the measurement tip. A ceramic filter, in conjunction with a deflecting plate, is used to keep the measuring face of the probe free of particulate matter.

Care must be taken with the cell because potassium sulfate is soluble in water. This requires that the electrolyte be at an elevated temperature. A temperature of 800°C is normally maintained. If the probe is left unheated in a saturated stack or in the rain, the potassium sulfate pellet may have to be replaced. Another concern with this technique is that an oxygen monitor must be used in conjunction with the $SO_2$ analyzer. This may not actually be a drawback, because it is necessary to measure $O_2$ in most combustion source applications to compute emission rates.

### A Polarographic Sensor for Measuring $SO_2$ and NO

Land Combustion Corporation has developed liquid electrochemical analyzers designed to measure flue gases on a point in-situ probe (Greaves 1989). Systems have been developed to measure both $SO_2$ and NO and NO only. They do not require an auxiliary oxygen measurement.

The analyzers utilize a polarographic technique, where the diffusion of the different gas species to the surface of an electrode is controlled, as in the extractive polarographic analyzers previously discussed. Here, a porous membrane is placed between the flue gas and an electrocatalytic electrode. The electrode material is sputtered onto the membrane, to provide a relatively porous surface that will allow molecules to react at the surface, pass into the electrolyte, and pass to the reference electrode. The reacting molecules generate a cell current that is proportional to its concentration in the flue gas (Fick's law of diffusion).

In the $SO_2$/NO instrument, as in laboratory polarographic analyzers, the type of molecules that react at the measuring electrode can be controlled by applying a voltage difference between the measuring electrode and the reference electrode. In order to measure both $SO_2$ and NO, the measuring electrode is divided into two sections and a different voltage potential is applied to each one. Because the polarograms of $SO_2$ and NO overlap (plotted as cell current vs. applied voltage), two simultaneous equations obtained for the differing applied potentials are solved for the pollutant gas concentrations. As with other point in-situ systems, the sensor can be flooded with calibration gas to check system performance.

### Zirconium Oxide Sensor for Measuring $O_2$

This discussion of point in-situ monitoring systems is not complete without providing some additional details about the zirconium oxide $O_2$ sensor in-situ analyzer. The $ZrO_2$ electrocatalytic technique has already been discussed with respect to its application in extractive systems. The technique's greatest popularity, however, has been in its application to in-situ

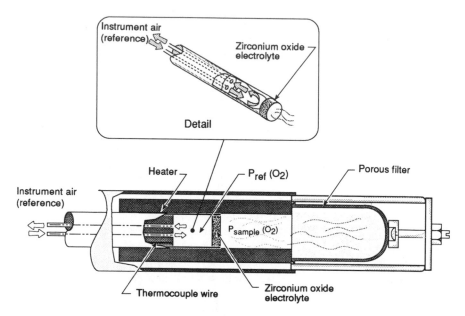

**FIGURE 6-14.**    Probe configuration of a zirconium oxide ($ZrO_2$) in-situ oxygen analyzer.

measurements. The method is used extensively for boiler control and monitoring, with over 20,000 units sold by one vendor alone.

The probe configuration of an in-situ zirconium oxide $O_2$ analyzer is shown in Figure 6-14. The zirconium oxide solid electrolyte separates the sample gas from reference gas. A heater surrounds the measurement cell to maintain a constant temperature of 850°C (1550°F), which enables the zirconium oxide to serve as a conduit for oxygen. The standard point in-situ monitor features, ceramic filter and probe calibration, are also incorporated in this system.

The zirconium oxide in-situ analyzer has shown remarkable success in flue-gas monitoring. Two factors contributing to this success are (1) the stability of $ZrO_2$ and (2) its ability to operate at an elevated temperature. The system was initially developed by a division of Westinghouse (now Rosemount) and has been widely copied. Today, the performance of the commercially available in-situ $O_2$ analyzers depends on the experience and quality control capabilities of the manufacturer, rather than on the method itself.

### *Advantages and Limitations of Point In-Situ Systems*
The idea of purchasing a box that can sit on a stack and deliver emissions data to the control room with just an electrical cable is very attractive to

a plant operator. Monitoring emissions without the necessity of conditioning the flue gas led to the original development of in-situ systems. In proper applications, oxygen and $SO_2$ can be measured reliably with point in-situ systems. In some cases, however, problems may occur.

Because the instrumentation is stack- or duct-mounted, the environment at the mounting location is extremely important. In situations where stack or duct vibration is present, point in-situ monitors may have problems operating, particularly the electro-optical systems. Vibration can loosen or misalign optical components (as with the path in-situ systems) and can loosen circuit boards and electronic components. For long probes, vibrations combined with varying flue-gas velocities may cause oscillations in the probe itself, leading to structural damage.

High ambient temperatures or widely fluctuating temperatures can also affect stack-mounted electronics and optics. Acid gas in the ambient air can affect poorly constructed instrument housings, and condensed acid gases in the flue gas can rapidly corrode the sensing probes. The use of corrosion-resistant alloys may be necessary in some applications.

Particulate matter may be a problem in other situations. Ceramic filters do help reduce the level of particulate matter entering the measuring cavity of a point monitor. However, submicron particles may still enter. The system must be able to compensate for such a problem (as most do). If the ceramic thimble should become plugged with particulate matter or condensed materials, flue gas will no longer diffuse into the cavity, a problem that should become evident during calibration. However, if the problem occurs frequently, the maintenance time required to clean or replace the filter may become excessive. Although blowback systems have been applied in some cases, they have not always been successful for ceramic thimbles. Deflection plates or bars or deflector sheaths are commonly used to reduce the impaction of particulate matter onto the filter.

It must also be noted that measurements are conducted at stack temperature and stack pressure. To calculate emissions with reference to standard conditions (20°C and 1 atm), the stack temperature and pressure must be known. In most cases, an assumed value for the stack pressure is used; however, the stack temperature is a significant factor in performing a correction to stack conditions. For this reason, most point in-situ systems also incorporate a thermocouple. A malfunctioning thermocouple can affect the corrected readings. Because temperature readings generally are not checked during a calibration or an audit, discrepancies between audit gas measurements and the emission measurements may, in some cases, be due to malfunctioning temperature compensation systems (Peeler 1981).

The capability of performing a cylinder gas probe calibration is a significant feature of the point in-situ analyzer. By injecting audit gas directly into the measurement cavity, an excellent assessment can be obtained of the instrument performance. Chapter 10 discusses this procedure further.

## FLUE-GAS VELOCITY (FLOW) MONITORS

Emission standards that specify the continuous monitoring of pollutant mass emission rates imply that velocity monitors are to be part of the CEM system. This can be seen by examining the following relation:

$$\text{pmr}_s = c_s Q_s \tag{6-11}$$

$$Q_s = A_s v_s \tag{6-12}$$

where  $\text{pmr}_s$ = pollutant mass rate (in kilograms per hour, pounds per hour, or tons per year)

$c_s$ = pollutant concentration (in grams per dry standard cubic meter or pounds per cubic foot)

$A_s$ = stack or duct area (in cubic meters or cubic feet)

$v_s$ = stack-gas velocity (in meters per second or feet per second)

with the application of appropriate conversion factors (e.g., seconds to hours, etc.).

The first U.S. federal requirement for continuous pollutant mass rate determinations appeared with the promulgation of Subpart LLL (Standards of Performance for Onshore Natural Gas Processing: $SO_2$ Emissions) in 40 FR 40160 October 1, 1985. Continuous rate monitoring requirements are more common in Europe, but have also begun to appear in U.S. state permits and other regulations. However, the allowance trading policy of the 1990 U.S. Clean Air Act Amendments—Title IV provides the greatest impetus for application of velocity monitors for the measurement of emission rates in the United States.

A number of techniques are currently available to monitor flue-gas volumetric flow rate. Although techniques that correlate gas velocity with parameters such as steam flow rate, combustion stoichiometry, or fan horsepower have been used for this purpose, they may be neither particularly accurate nor practical (Richards 1989). Monitors used to measure flue-gas velocity are inherently in-situ monitors, because a dynamic gas measurement must be made. The methods vary in complexity from the use of pitot tubes to transmitting ultrasonic signals across the stack (Table

TABLE 6-1     **Flow-Monitoring Techniques**

| Technique | Instrumentation or Sensor |
|---|---|
| differential pressure sensing | Head meters, pitot tube, averaging probe |
| | Fluidic sensors |
| Thermal sensing | Heated sensors |
| Acoustic velocimetry | Ultrasonic transducers |

6-1). Other instrumentation, such as orifice meters, venturi meters, vane anemometers, and flow tubes, are more appropriate to air-handling ducts or specialized gas streams free of particulate matter.

To obtain a value for the volumetric flow, the differential pressure and thermal techniques require the measurement of stack temperature and/or flue-gas density. An averaging probe can be designed to measure at centers of equal areas across a stack or duct, whereas the ultrasonic method measures velocity on a line average. The other methods measure at one point only, although arrays of sensors can be used to measure at centers of equal areas on a cross section. This discussion of flow-rate monitoring techniques is not intended to be exhaustive of all methods currently available. It emphasizes those that have been or are being used for monitoring flow in industrial or utility stacks or flues.

## Differential Pressure Sensing

### The Pitot Tube

Pitot tubes have been used traditionally to measure stack-gas volumetric flow rate and are specified as the EPA reference for measuring flue-gas velocity. A pitot tube consists of two tubes, one facing the direction of flow of the gas, to measure an impact pressure, and the other tube either perpendicular to the flow or in the direction opposite the flow, to measure the static or wake pressure. The pressure differential between the stagnation pressure and the wake pressure is the "velocity pressure" $\Delta p$ and is measured using a manometer, Magnehelic gauge, or pressure transducer. The velocity pressure $\Delta p$ is related to the flue-gas velocity by the pitot tube equation:

$$v_s = K_p C_p \sqrt{\frac{T_s \, \Delta p}{P_s M_s}} \tag{6-13}$$

where   $v_s$ = velocity of the gas
$K_p$ = dimensional constant
$C_p$ = pitot tube calibration coefficient
$T_s$ = absolute temperature of the gas
$P_s$ = absolute pressure of the stack gas
$M_s$ = molecular weight of the stack gas

Note that a number of parameters beside the directly measured $\Delta p$ must be either assumed, calculated, or otherwise determined, to obtain the velocity.

A number of devices have been developed to take advantage of these principles. The simplest is the type-S (Stausscheibe) pitot tube specified in U.S. EPA Reference Method 2. The stagnation and wake pressures can be monitored continuously using pressure transducers (capacitance-type or other) and, by using a thermocouple to monitor stack temperatures, the velocity can be calculated (stack-gas molecular weight is estimated). Plugging of tubes can be avoided by periodically blowing high-pressure air through the tubes (Rollins 1977).

A pitot tube obviously measures at only one point, although multiple tubes can be used to average a flow distribution across the stack or duct. The impact and static tubes can feed into separate, common chambers where the pressures equilibrate to give an average value. However, a simpler averaging technique is employed in the averaging probe.

### Averaging Probes

Averaging probes are a modified form of pitot tube, having four or more ports in a pipe, located at the traverse points corresponding to the centers of equal areas of the stack cross section. These ports face the direction of flow and give an average stagnation pressure over the stack diameter. The static pressure is averaged using ports located behind the high-pressure ports. Because the port locations will be different for each installation, stack dimensions must be carefully specified before the probe is constructed.

An averaging probe averages only on one diameter. If two probes were installed perpendicular to each other, the flow would be more completely characterized. The accuracy of an averaging probe is dependent, as is that of the pitot tube, on the constancy of its calibration coefficient ($C_p$) and assumptions associated with the stack-gas density (molecular weight or composition and temperature).

The pitot tube and averaging probe are less sensitive to low flow rates than to high flow rates because low pressure differentials ($\Delta p$) are difficult to measure accurately. Also, agglomerating particulate matter, acid gases, and moisture droplets may cause system failures. However, the use of blow-back techniques can, again, increase availability. There is also a question of cross-flow existing in averaging probes due to different port pressures (Ginesi and Grebe 1987), but the severity of such a problem would become evident when certifying the installation.

### Thermal Sensing Systems

Thermal sensing instruments are based on the transfer of heat from a heated body to the flowing gas. This requires the use of two sensors, one heated, the other unheated. In a typical configuration (Figure 6-15), platinum resistance wire is wound on ceramic cylinders, which are then protected by a stainless steel tube. The longer sensor is heated (the velocity sensor); the shorter one (temperature sensor) is not heated. These two resistance elements are connected to a bridge circuit that maintains the temperature of the heated sensor. As the moving stack gas cools the sensor, the current through the element is increased to keep the temperature constant. This current (and the resultant voltage signal that is generated) is related to the heat loss from the sensor. The unheated sensor is used to compensate for temperature changes in the stack gas.

In another configuration, the heated and unheated resistance elements are combined on a single tip. The tip is glass-coated and is typically applied to monitoring flow in noncorrosive atmospheres. A metal-clad

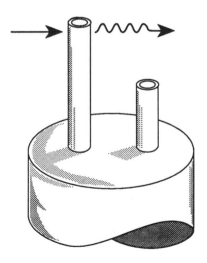

**FIGURE 6-15.**    A thermal probe sensor.

sensor is used in industrial process control and stack-gas monitoring. The larger-diameter sensors are said to minimize the buildup of particulate matter. Particulate buildup should be avoided because it affects the sensor calibration.

Thermal sensing systems differ from the differential-pressure-type instruments in a fundamental manner in that they measure mass flow directly, not volumetric flow. The rate of cooling of the heated sensor is dependent upon the thermal conductivity of the stack gas, which is dependent upon the gas viscosity and specific heat. Gas viscosity in turn is dependent upon velocity over a given path and on the gas density. As a net result of these effects, thermal sensing instruments produce an output that is proportional to mass flow and not volumetric flow:

$$\{\text{Signal}\} \propto \{\text{heat loss}\} \propto \rho v_s A_s \left[ (\text{kg/m}^3)(\text{m}^3/\text{h}) \text{ or } (\text{lb gas})/\text{ft}^3)(\text{ft}^3/\text{hr}) \right]$$

$$= \text{kilograms of gas per hour or pounds of gas per hour} \qquad (6\text{-}14)$$

where $\rho$ is the gas density

Note that this is the mass flow of the stack gas, not the mass flow of pollutant.

When knowledge of mass flow rates is desired (for example, in process control) the technique allows their determination without requiring a knowledge of the gas pressure and temperature. However, when velocity or volumetric flow is needed (cubic meters per hour or cubic feet of gas per hour), a knowledge of the gas density (pounds per cubic foot or grams per cubic centimeter) is required. Because the gas density is dependent on its composition, changes in stack-gas moisture content, $CO_2$ concentrations, and so on will affect the calibration of the instrument.

One advantage of the thermal systems is that multiple sensors can be combined easily in arrays to measure the average flow rate over the cross section of a duct or stack. Because each sensor makes an independent measurement, it is possible to monitor the gas flow distribution over the cross section.

## Acoustic Velocimetry

This method measures the time that it takes sound pulses to travel with the direction of flow of the stack gas and against the direction of flow of the stack gas. Although the technique has been utilized in monitoring liquid flow, one of the first applications to stack-gas monitoring met with only limited success (Brooks et al. 1975). With improvements in ultrasonic

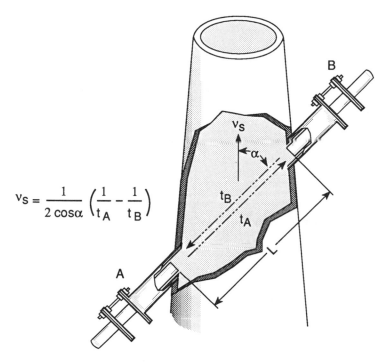

$$v_S = \frac{1}{2\cos\alpha}\left(\frac{1}{t_A} - \frac{1}{t_B}\right)$$

**FIGURE 6-16.**     Acoustic measurement of flue-gas velocity.

transducers and the advent of digital electronics, the technique has been successfully applied and has met U.S. EPA performance specifications and the suitability criteria for stack-gas monitoring in the Federal Republic of Germany.

In this method, ultrasonic pulses in the range of 50 kHz are transmitted both upstream and downstream of the flow (Figure 6-16). Two transceivers are located opposite each other on the stack at an angle of typically 45°. In each transceiver, a piezoelectric transducer transmits ultrasonic pulses over the path $\ell$ to the opposite transceiver. The transducers both convert electric signals to acoustic signals and acoustic signals to electric signals. The speed at which the pulse crosses the stack is dependent upon whether it is going with or against the flow:

1.  Speed in direction of flow,

$$v_A = \frac{\ell}{t_A} = c + v_s \cos\theta$$

where $\ell$ = path between $A$ and $B$
$\quad t_A$ = forward transit time from $A$ to $B$
$\quad v_A$ = speed from $A$ to $B$
$\quad c$ = speed of sound
$\quad v_s$ = stack gas velocity
$\quad \theta$ = angle between stack and path $\ell$

2. Speed against direction of flow,

$$v_B = \frac{\ell}{t_B} = c - v_s \cos\theta$$

where $t_B$ = transit time from $B$ to $A$
$\quad v_B$ = speed from transceiver $B$ to transceiver $A$

Note that $\pm v_s \cos\theta$ is the vector contribution, in the $\ell$ direction, of the stack-gas velocity to the transit of the pulse.

If these two equations are subtracted, the stack-gas velocity can be obtained.

$$v_s = \frac{\ell}{2\cos\theta}\left[\frac{1}{t_A} - \frac{1}{t_B}\right] \tag{6-15}$$

This expression is independent of the sound velocity $c$ and, when used with the stack cross-sectional area, gives the volumetric flow rate in *actual* cubic feet per minute. Note that the expression is independent of other gas properties such as density, pressure, or temperature.

If the expressions for the forward and reverse velocities are added, the speed of sound $c$ can be obtained:

$$c = \frac{\ell}{2}\left[\frac{t_A + t_B}{t_A t_B}\right]$$

Because the speed of sound is temperature-dependent by the relation

$$c = KT^{1/2} \quad \text{(where } K \text{ is a constant)}$$

the stack gas temperature is then

$$T = \left(\frac{c}{K'}\right)^2 \quad \text{(where } K' = 1/K^2\text{)}$$

In this case, $c$ is dependent upon stack-gas composition and the specific heats of the stack gases. However, correction factors can be applied with reasonable success.

A feature unique to the technique is that it provides a mechanism for an internal check of the system. An internal zero can be obtained by electronically substituting the signal going with the flow for the one going against the flow. This should result in $t_A = t_B$ and $v_s = 0$. An upscale system check can be obtained by introducing a known delay in the tone pulses and monitoring the delay

Because the instrument is a cross-stack path system, it is not subject to the corrosion and particulate fouling problems that may affect the in-stack, insertion-type probes. However, particulate matter can foul the transceivers, so purge air is blown directly through them.

## EXTRACTIVE SYSTEMS VERSUS IN-SITU SYSTEMS

A continuing matter of discussion revolves around determining which system is better: extractive or in-situ. The simplest answer to the question is, again, there is *no* "best" system because CEM systems are all application-dependent. In fact, CEM system data quality is not as dependent upon the type of CEM system installed as it is dependent on the performance of maintenance at a level sufficient to maintain data quality. However, the most accurate CEM systems will utilize methods that have the ability to measure the gas of interest without interference from other gases or materials found in the gas stream.

With these qualifications in mind, some of the principal differences between the methods are summarized in Table 6-2. There are also many differences between the types of CEM systems available commercially. As with any instrumentation, some systems are antiquated in design, others are poorly constructed, and others may require excessive amounts of maintenance (whether extractive or in-situ). Vendors similarly vary in quality and capability. Some CEM system vendors maintain well-managed and well-developed organizations, whereas others are made up of a few individuals without adequate service staffs or financial resources. As in purchasing any product, it is left to the astuteness of the buyer to choose the best system and vendor for his or her application.

**TABLE 6-2    Comparison of Features of Extractive and In-Situ Systems**

| Extractive Systems | In-Situ Systems | |
| | Path | Point |
| --- | --- | --- |
| Can sample from point of average concentration | Linearly average concentrations | Can sample from point of average concentration |
| Can time-share analyzers | Cannot time-share analyzers | |
| Analyzers readily located in environmentally controlled shelters or rooms | Analyzer subject to ambient environmental conditions | |
| Have large number of maintainable system components | Have fewer maintainable system components | |
| System components may be relatively easy to repair (pumps, valves) | System components may be relatively difficult to repair (optics, electronics) | |
| Use cylinder gases for calibration | May use internal gas cells for calibration | May use internal gas cells for calibration |
| Can easily audit with cylinder gas | Cylinder gas audits may require special attachments | Can easily audit with cylinder gas |
| Do not require temperature compensation | Require temperature compensation | |
| Sample gas filtered, conditioned to standard standard temperature and pressure | High stack temperatures, high levels of particulate matter, and sticky particulate matter may affect performance | |
| May alter sample | Do not alter sample | Do not alter sample |
| Response time depends on sample line length | Response time depends on analyzer response, not sampling | |
| System problems readily solved on-site | System problems may be difficult to solve on-site (e.g., electro-optical problems at stack height) | |
| Maintenance may not require special training | Maintenance may require higher levels of training | |

### References

Bambeck, R. J., and Huettemeyer, H. F. 1974. Operating experience with in-situ power plant stack monitors. Paper presented at the Air Pollution Control Association Meeting, Denver. Paper 74-109.

Biermann, H. S., and Winer, A. M. 1990. Recent improvements in the design and operation of a differential optical absorption spectrometer for in situ measurement of gaseous air pollutants. Paper presented at the Air Pollution Control Association Meeting, Pittsburgh. Paper 90-87.2.

Brooks, E. F., Beder, E. C., Flegal, C. A., Luciani, D. J., and Williams, R. 1975. *Continuous Measurement of Total Gas Flowrate from Stationary Sources*. EPA 650/2-75-020.

Brooks, E. F., and Williams, R. L. 1976. *Flow and Gas Sampling Manual*. EPA-600/2-76-203.

Dec, D. A. 1986. A new instrument for measuring water vapor concentration. Paper presented at the Air Pollution Control Association Meeting, Minneapolis. Paper 86-71.4.

Fyock, D. H., Blau, H. H., Fasci, E. W., Kebabian, P. L., Kruse, J. R., and Weiser, R. 1975. Test of the environmental research and technology stack gas analyzer at the Conemaugh generating station. Paper presented at the Air Pollution Control Association Meeting, Boston. Paper 75-60.6.

Ginesi, D., and Grebe, G. 1987. Flow—a performance review. *Chem. Eng.* 6:102–118.

Greaves, K. 1989. New detector design for simultaneous monitoring of $SO_2$ and $NO_x$. In *Advances in Instrumentation and Control*, Vol. 44, *Proceedings of the ISA/89 International Conference & Exhibit, Philadelphia*. Instrument Society of America, Research Triangle Park, NC, pp. 379–391.

Hager, R. N. 1973. Derivative spectroscopy with emphasis on trace gas analysis. *Anal. Chem.* 45(13):1131A–1138A.

Hager, R. N., and Anderson, R. C. 1970. Theory of the derivative spectrometer. *J. Opt. Soc. Amer.* 60(11):1444–1449.

Herget, W. F., Jahnke, J. A., Burch, D. E., and Gryvnak, D. A. 1976. Infrared gas-filter correlation instrument for in-situ measurement of gaseous pollutant concentrations. *Applied Optics* 15:1222–1228.

Jahnke, J. A., and Aldina, G. J. 1979. *Continuous Air Pollution Source Monitoring Systems—Handbook*. EPA 625/6-798-005.

Jones, J. E. 1985a. Evaluation of four $SO_2$ emissions monitors downstream of utility boiler wet scrubber. In *Transactions—Continuous Emission Monitoring: Advances and Issues* (J. A. Jahnke, Ed.). Air Pollution Control Association, Pittsburgh, pp. 414–430.

Jones, J. E. 1985b. $SO_2$ analyzer continuously helps power plant comply. *Pollution Engineering* 6:38–40.

Karlsson, R. 1990. Environmental control using long path measurements. In *Measurement of Toxic and Related Air Pollutants—Proceedings of the 1990 EPA/A & WMA International Symposium, Raleigh*. Air and Waste Management Association, Pittsburgh, pp. 675–784.

Lord, H. C. 1976. Verification of in-situ emission analyzer data. In *Calibration in Air Monitoring*, ASTM STP 598. American Society for Testing and Materials, Philadelphia, pp. 107–117.

Lord, H. C. 1982. In-situ monitoring: Operational experience bolsters reliability. *Power* 5:88–95.

Nation, D. K. 1981. Continuous emission monitoring experience at Colstrip Units 1 & 2. In *Proceedings—Specialty Conference on Continuous Emission Monitoring: Design, Operation and Experience*. Air Pollution Control Association, Pittsburgh, pp. 25–38.

Nelson, R. L. 1984. Solid electrolyte reduces cost of in-situ $SO_2$ monitoring. *Pollution Engineering* 1:26–27.

Nelson, R. L. 1985. EPA $SO_2$ emission monitoring is simplified with new in-situ solid electrolyte analyzer. Paper presented at the Air Pollution Control Association Meeting, Detroit. Paper 85-51.8.

Peeler, J. W. 1981. Continuous emission monitor performance evaluation and quality assurance project: Possum Point Power Station, Unit #5, Virginia Electric and Power Company. In *Proceedings—Specialty Conference on Continuous Emission Monitoring: Design, Operation and Experience*. Air Pollution Control Association, Pittsburgh, pp. 221–232.

Richards, J. 1989. Utility boiler parameter monitoring for determination of flue gas flow rates. Unpublished contract report, Richards Engineering (Raleigh, NC).

Rollins, R. 1977. A continuous monitoring system for sulfur dioxide mass emissions from stationary sources. Paper presented at the Air Pollution Control Association Meeting, Toronto. Paper 77-27.5.

Schuck, C. M. 1981. Southern California Edison Company experience with in-situ and extractive $NO_x$ monitors. In *Proceedings—Specialty Conference on Continuous Emission Monitoring: Design, Operation and Experience*. Air Pollution Control Association, Pittsburgh, pp. 48–61.

Vaughan, W. M. 1991. Remote sensing terminology. *J. Air & Waste Mgmt. Assoc.* 41:1489–1493.

Williams, D. T., and Palm, C. S. 1974. Evaluation of second derivative spectroscopy for monitoring toxic air pollutants. National Technical Information Service Report No. SAM-TR-74-19, September 1974.

**Bibliography**

ASME. 1959. *Fluid Meters—Their Theory and Application*. The American Society of Mechanical Engineers, New York.

Baker, W. C., and Pouchot, J. F. 1983. The measurement of gas flow—Parts I and II. *J. Air Pollut. Control Assoc.* 33(1):66–72 and 33(2):156–162.

Chadbourne, J. F. 1979. Continuous monitoring of gaseous emissions on cement kilns. Paper presented at the Air Pollution Control Association Meeting, Cincinnati. Paper 79-22.1.

Federal Ministry for the Environment, Nature Conservation and Nuclear Safety. 1988. *Air Pollution Control Manual of Continuous Emission Monitoring*. Bonn, Germany.

Hager, R. N., and Anderson, R. C., Minnich, T. R., Scotto, R. L., Kagann, R. H., and Simpson, O. A. 1990. Air monitoring—optical remote sensors ready to tackle Superfund RCRA emissions monitoring tasks. *Hazmat World* 3(5):42–59.

Klompstra, T. A. 1990. A comparison of extractive and in-situ technology. In *Proceedings—Specialty Conference on Continuous Emission Monitoring: Present and Future Applications.* Air and Waste Management Association, Pittsburgh, pp. 84–92.

Knoepke, J. 1977. Tracer-gas system determines flow volume of flue gases, *Chem. Eng.* 1:91–94.

Lindenberg, S. P. 1981. Continuous emission monitor performance specification testing and operation at Coal Creek Station. In *Proceedings—Specialty Conference on Continuous Emission Monitoring: Design, Operation and Experience.* Air Pollution Control Association, Pittsburgh, pp. 70–80.

Maeda, M., and Shunya, N. 1990. A new combustion sensor for exhaust gases from boilers, furnaces and automobile engines. In *Advances in Instrumentation and Control,* Vol. 44, *Proceedings of the ISA/89 International Conference & Exhibit—Philadelphia.* Instrument Society of America, Research Triangle Park, NC, pp. 469–477.

Mullowney, R. L., and Kirchner, D. L. 1981. In-situ gas monitor experience at CIPS. In *Proceedings—Specialty Conference on Continuous Emission Monitoring: Design, Operation and Experience.* Air Pollution Control Association, Pittsburgh, pp. 143–150.

Opalinski, M., and Townsend, A. 1986. CEM system at the Seminole Power Plant to comply with Subpart Da performance specifications and requirements. In *Transactions—Continuous Emission Monitoring: Advances and Issues.* Air Pollution Control Association, Pittsburgh, pp. 31–43.

Polhemus, C. 1976. The design and performance of a spectrometer for in-situ measurement of $SO_2$ and NO. In *ISA Analysis Instrumentation Proceedings,* Vol. 14. Instrument Society of America, Research Triangle Park, NC.

Polhemus, C. 1981. Performance evaluation testing of compliance monitors. In *Proceedings—Specialty Conference on Continuous Emission Monitoring: Design, Operation and Experience.* Air Pollution Control Association, Pittsburgh, pp. 151–164.

Polhemus, C., and Hudson, A. 1976. A performance analysis of Lear Siegler's in-situ $SO_2$/NO monitor. Paper presented at the Air Pollution Control Association Meeting, Portland. Paper 76-35.5.

Pruce, L. M. 1980. Maximize boiler efficiency and save fuel with CO monitoring. *Power.* 1:70–72.

Reason, J. 1981. When it pays to monitor flue-gas CO. *Power.* 8:37–43.

Scherrer, C. R. 1986. CEM experience on a high sulfur wet FGD system. In *Transactions—Continuous Emission Monitoring: Advances and Issues.* Air Pollution Control Association, Pittsburgh, pp. 9–16.

Traina, J. E. 1985. Reliability of a continuous stack flow monitor and application in improved process/control equipment operation. In *Transactions—Continuous Emission Monitoring: Advances and Issues* (J. A. Jahnke, Ed.). Air Pollution Control Association, Pittsburgh, pp. 261–267.

Traina, J. E. 1990. $SO_2$ mass emissions measured by new CEM technique. In *Proceedings—Specialty Conference on Continuous Emission Monitoring: Present and Future Applications.* Air and Waste Management Association, Pittsburgh, pp. 126–135.

U.S. Government. 1990. *Clean Air Act Amendments of 1990—Conference Report to Accompany S. 1630*. Superintendent of Documents, U.S. Government Printing Office, Washington, DC.

Webb, R. C., and Opel, A. E. 1979. True on-line zero and span determination for a continuous emission monitoring system. Paper presented at the Air Pollution Control Association Meeting, Cincinnati. Paper 79-35.4.

Wirth, R. G. 1974. Fluidic sensors to measure velocities under adverse conditions. In *Air Quality Instrumentation*, Vol. 2. Instrument Society of America, Pittsburgh.

# 7

## Transmissometers

A transmissometer measures the transmittance of light that passes through a flue gas. The transmissometer (for "transmission meter") is also called an opacity monitor—the two terms are interchangeable. Transmissometer measurements are performed in-situ and can be used to monitor the performance of air pollution control equipment or compliance with source opacity standards. However, the intrinsic property of in-stack opacity does not always form the basis for emissions standards, as does pollutant gas concentration. For example, in the United States, federal opacity standards are based on visual determinations made by trained observers [EPA Reference Method 9 (U.S. EPA 1991a)]. In Europe, opacity monitor data is commonly correlated with manual stack test data to check compliance with mass rate standards. The direct use of in-stack opacity measurements is not precluded for incorporation in emission standards, and environmental control agencies have, in fact, adopted regulations in this form.

The way in which a transmissometer is to be used can affect its design. If the instrument data are intended to correlate with the visible emissions observer, the transmissometer should determine opacity using visible (photopic) light. If the data are to be correlated with particulate mass rate, red or infrared light may be more appropriate. Other design features can be incorporated in a transmissometer to enhance overall system performance. This chapter discusses design criteria for the transmissometer and shows how they are applied in a number of commercially available systems. Because of the importance of optical density calculations in transmissometer applications, calculation techniques are also reviewed.

157

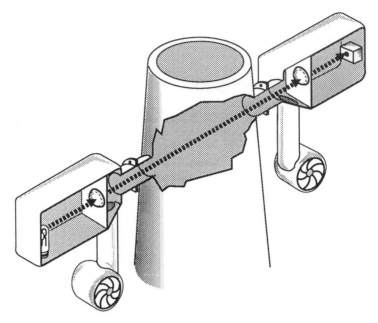

**FIGURE 7-1.**    Single-pass transmissometer system.

## BASIC COMPONENTS OF TRANSMISSOMETERS

In a transmissometer, the optical system is designed so that the transmittance is determined from two principal measurements: the measurement of $I_0$ (a reference measurement of light intensity) and $I$ (the intensity of the light beam measured after the light passes through the flue gas). The ratio of the two intensities $(I/I_0)$ is the fractional transmittance (Tr), which, as seen from Equation (4-6), is equal to $(1.0 - \text{Op})$ where Op is the fractional opacity.

As with in-situ path gas monitoring systems discussed in the previous chapter, transmissometers may be constructed using either a single-pass design (Figure 7-1) or a double-pass design (Figure 7-2). In the single-pass design, a lamp projects a beam of light across the stack or duct leading to the stack, and the amount of light transmitted through the flue gas is sensed by a detector. Such instruments can be made rather inexpensively and are often used as bag house monitors. However, they generally do not satisfy regulatory design criteria.

The double-pass system shown in Figure 7-2 houses both the light source and light detector in one unit, the transceiver assembly. By reflecting the projected light from a retroreflector mirror housed on the opposite

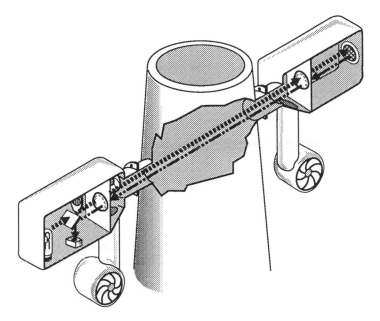

**FIGURE 7-2.**    Double-pass transmissometer system.

side of the stack, systems can be easily designed to check the electronic circuitry, including the lamp and photodetector, as part of the operating procedure. Most transmissometer systems include some type of air purging system or blower to keep the optical windows clean. In the case of stacks with a positive static pressure, the purging system must be efficient or the windows will become dirty, leading to spuriously high readings. Shutters or flaps are often used to seal off the flow gas from the windows in case of blower failure.

## DESIGN SPECIFICATIONS

Gas monitoring systems can be checked readily, using manual or instrumental reference test methods. However, for opacity measurements, there is no scientifically independent or reasonably convenient method of checking the accuracy of stack-mounted transmissometers. In the United States, design specifications therefore have been established to provide a guide to the development of opacity monitoring instruments that generate data that can be correlated with visible emissions observations. Design specifications also set standards of performance so that measurements from source to source will be made as uniformly as possible. Transmissometer design specifications are given in U.S. EPA Performance Specification 1 for the following transmissometer features (U.S. EPA 1983, 1991c):

- spectral response (peak and mean)
- angle of view
- angle of projection
- optical alignment sight
- simulated zero and upscale calibration system
- external calibration filter access
- automatic zero compensation indication
- specifications for instruments with slotted tubes

### Spectral Response

A transmissometer designed for regulatory applications (where the data are to correspond to visible emissions observations) must respond to light in the visible, photopic, range of the electromagnetic spectrum. The specification (U.S. EPA 1991c) states: "The peak and mean spectral responses must occur between 500 and 600 nm. The response at any wavelength below 400 nm or above 700 nm shall be less than 10% of the peak spectral response."

Figure 7-3 shows a possible spectral response curve for a transmissometer. The peak response is the wavelength where the transmissometer is the most sensitive and is the wavelength that corresponds to the highest value

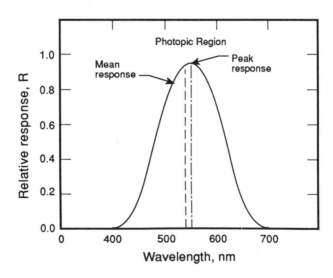

**FIGURE 7-3.**    Example spectral response curve for a transmissometer.

of the relative instrument response on the curve. The mean spectral response is the wavelength where the transmissometer shows its mean, or average, sensitivity.

There are three basic reasons for specifying transmissometers to measure in the visible range (400–700 nm) of the spectrum.

1. If an instrument is designed to monitor light attenuation in this region of the spectrum, the measurements can be compared to the data obtained by emissions observers. This comparability is useful for both scientific and regulatory cross-checks of opacity monitoring methods.
2. Water vapor and $CO_2$ absorb light energy in the near infrared range of the electromagnetic spectrum (Figure 4-2). If a transmissometer projects light in this range, water vapor or $CO_2$ can absorb some of the light energy by molecular absorption processes. This would result in a higher opacity reading than that obtained using an instrument that projects light at shorter wavelengths in the visible range. Tungsten lamps emit light in the infrared region, but can still be used in transmissometer systems if appropriate optical filter–detector combinations are employed to eliminate the detection of light at the higher wavelengths.
3. Light scattering is very dependent on the size of the particle relative to the wavelength of the light impinging on it (as discussed in Chapter 4). For example, light at 450 nm has a maximum Mie scattering coefficient for particles in a range of 0.2 $\mu$m diameter, whereas light at 1000 nm has a smaller scattering coefficient for 0.2-$\mu$m-diameter particles and exhibits maximum scattering for particles having sizes 0.5 $\mu$m and larger (Cashdollar, Lee, and Singer 1979). In industrial plants controlled with bag houses and electrostatic precipitators, most of the larger particles are removed. Because the stack-exit plume opacity results from the scattering of visible light from the uncaptured small particles, it is essential that monitors be designed to measure the loss of transmittance due to small-particle scattering.

### Angle of Projection and Angle of View

In Chapter 4, it was noted that light will scatter in many different directions after it impinges on a particle. If a transmissometer had a large detector or a lens that gathered light over a wide angle, it might measure the light that would ordinarily be lost to its field of view. As a result, the measured opacity would be lower than it should be, or the transmittance would be too high; the detector would sense the scattered light, which would, therefore, be counted as transmitted light.

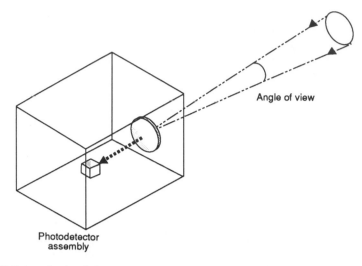

Angle of view

Photodetector
assembly

**FIGURE 7-4.**    Angle of view.

Similarly, if the lamp and lens system projected a cone of light at a wide angle, light could be scattered into the view of the detector instead of out of the view of the detector. This again would increase the measured transmittance and decrease the opacity. Because of the importance of this effect, design specifications have been established for the observation cone of the detector and for the spread of the cone of projected light. These specifications are expressed as (1) the angle of view and (2) the angle of projection. Definitions given for monitor assemblies in Performance Specification 1 (U.S. EPA 1991c) are as follows:

> **Angle of view**   The angle that contains all of the radiation detected by the photodetector assembly of the analyzer at a level greater than 2.5% of the peak detector response
>
> **Angle of projection**   The angle that contains all of the radiation projected from the lamp assembly of the analyzer at a level of greater than 2.5% of the peak illuminance

Measurements are normally made from the transceiver window or from the windows or lenses of the transmitter and receiver units of the monitor. (Figure 7-4 shows the angle of view and Figure 7-5 the angle of projection.) The U.S. EPA requires that these angles are to be no greater than 5°.

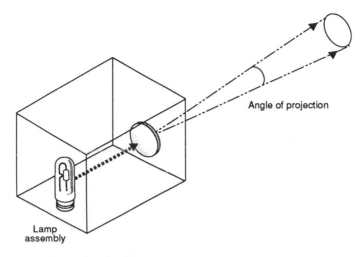

Angle of projection

Lamp
assembly

**FIGURE 7-5.**    Angle of projection.

## Optical Alignment Sight

A well-designed transmissometer system will incorporate some method to indicate optical misalignment. In a stack- or duct-mounted double-pass monitor, temperature changes, vibration, or poor maintenance procedures may cause the retroreflector assembly to shift with respect to the transceiver. As a result, not all of the projected light will hit the reflector. Less light will be returned to the detector, and the opacity will read higher than it actually is in the stack. A satisfactory alignment sight will indicate a misalignment that leads to an error of 2% opacity or greater at a monitor pathlength of 8 m.

## Simulated Zero- and Upscale-Calibration Systems

Commercial transmissometers have been designed to produce a "simulated zero" value and an upscale attenuation reading. The term "simulated" is used because most of the techniques applied do not actually produce a zero measurement across a stack or duct.

This simulation process is required to check the optical and electronic components of the instrument on a periodic basis. All of the double-pass monitors provide some method of simulating a zero opacity, either by using a "zero mirror," or by using two different upscale attenuators. The instrument need not actually produce a zero value, but may generate a 0–10% opacity as one point of what is essentially a two-point calibration requirement.

The upscale calibration system can be either a screen, a series of metal slats, an optical glass or gel filter, or a reflective system.

## Access to External Optics

A well-designed system will provide convenient access to the lenses exposed to the stack gas, so that they can be cleaned. Even well-designed blower systems cannot keep these surfaces clean indefinitely, so the monitor housings must be designed with this point in mind. Removing the entire instrument for cleaning is not appropriate because this would be a cumbersome procedure for what is intended as periodic maintenance. Neither should it be necessary to realign the unit after window cleaning.

## Automatic Zero Compensation Indicator

It is relatively easy to design a transmissometer system to compensate automatically for the buildup of particulate matter on the exposed optical surfaces. Either through the use of an external zero mirror or extra detectors, this feature provides the instrument operator with a means of checking the cleanliness of the exposed windows. If there is an electronic means for checking the amount of dirt buildup, one generally can provide an electronic means of compensating for it—subtracting it from the stack readings.

Instruments with a zero compensation system must also provide a method to read the amount of compensation. The instrument designer should not "bury" what is, essentially, a correction factor in the system electronics. If it were buried, the operator would not be able to identify developing problems in the purge air system or unusual variations in soiling rates. The U.S. EPA design specifications allow compensation to extend up to 4% opacity. After this value is exceeded, the instrument system must indicate to the operator that the windows need cleaning. Compensation systems on double-pass units only correct for soiling on the transceiver window.

## Slotted Tube Requirements

A slotted tube is sometimes used to maintain the optical alignment of a transmissometer system. The tube, set between the transceiver–transmitter and retroreflector–receiver, provides a stable mounting support for the instrument assemblies located on opposite sides of the stack. However, slotted tubes must be designed carefully. Design criteria include the following:

- The length of the open, slotted portion of the tube must be equal to or greater than 90% of the distance between the stack or duct walls.
- The tubes must not interfere with the free flow of stack gas through the slots.
- Light reflections must be minimized.

### External Calibration Filter Access

Special audit devices are now being applied in quality assurance audit programs (Chapter 10) to check transmissometers. These devices enable independent neutral density filters to be used to check transceiver performance. The transmissometer design specifications recommend as an option design criterion that transmissometers be constructed in such a manner that an audit device can be attached to the transceiver.

### Design Specification Verification

Verification that transmissometers are property designed and meet required specifications is time-consuming and requires the use of special equipment. Because analyzers of a given model are generally designed and manufactured in the same manner, there is little purpose to checking each instrument. However, the U.S. EPA does require that one analyzer, randomly selected from each month's production, be tested according to design specification verification procedures given in Performance Specification 1. The purchaser of a transmissometer must then obtain a "Certificate of Conformance" that states that during the month the analyzer was produced, one analyzer underwent all verification tests. The certificate should be obtained from the instrument vendor and must include the results of each test performed on the sampled analyzer.

## TRANSMISSOMETER DESIGN

It was mentioned previously that transmissometer optical systems are designed so that two principal measurements can be made. These are the measurement for $I_0$ (the intensity of a reference light beam) and the measurement of $I$ (the intensity of a light beam that has passed through the flue gas). The ratio of the two intensities, of course, gives us the transmittance.

Three categories of methods are commonly used to obtain this ratio in commercially marketed transmissometers. These methods incorporate (1) optical choppers, (2) rotating reference wheels, and (3) optical fibers. In addition, a unique instrument, marketed by Rosemount, Inc., uses liquid-crystal components to obtain transmittance measurements without incorporating moving parts.

## Chopper Techniques

The most notable transmissometers that use a chopper to alternate between the measurement of $I_0$ and $I$ are two German instruments, one manufactured by Erwin Sick, Inc., and marketed by Land Combustion, Ltd., in the United States. The other is manufactured by Durag Corporation and marketed by Enviroplan in the United States.

### The Erwin Sick Transmissometer

The Erwin Sick instrument is a double-pass system and performs the transmittance measurement by using a "beam splitter" and a reflective chopper. These components are shown in Figure 7-6. Note from Figure 7-6 that the light projected from the lamp passes through two lenses, a light modulator, and then a beam splitter. A light passes through the beam splitter and then encounters a reflective chopper. The chopper bounces the beam back to the beam splitter, which then reflects the light down to a detector. The detector measures the incident light in this part of the instrument cycle as the $I_0$ value (reference zero value). The reflective chopper continuously rotates, so that it alternately reflects the light beam or allows it to pass out of the transceiver assembly to the retroreflector (Figure 7-6). When the light reaches the retroreflector, it returns to make a second pass through the gas and reenters the transmitter–receiver (transceiver).

On this return trip, the light again passes through the reflective chopper when the lobes are positioned properly. It then reflects from the surface of the beam splitter to be measured at the detector to give the value for $I$. The ratio $I/I_0$ then gives the transmittance, which can be converted easily into opacity or optical density.

Note also several other components of the system. The light modulator is used to code the transmitted light so that the detector will be insensitive to ambient light. The modulator has a number of holes, which essentially cause the light to pulse. The instrument detector and electronic system are designed to detect only the pulsed, or modulated, light. Unmodulated light that reaches the detector from the stack exit or stack openings will not be detected.

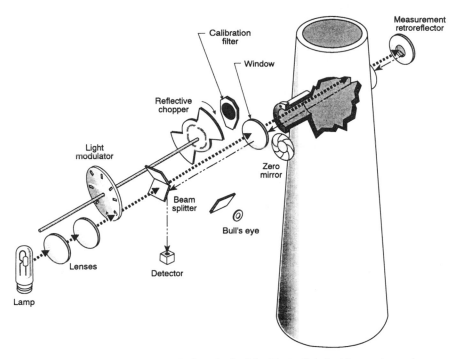

**FIGURE 7-6.**   Measurement mode ($I$ value) of the Erwin Sick double-pass transmissometer.

This transmissometer design also provides a calibration system for simulating a zero value and an upscale opacity. A mirror, called the zero reflector, is moved into the path of the light beam using a motor (Figure 7-7). This gives a simulated or "pseudozero" reading for the instrument. When the mirror is in this position, an upscale calibration attenuator can also be moved into the path of the light beam. This produces an opacity value that checks the electrical and optical components of the transceiver.

Because the zero reflector is located outside the window that seals the transceiver from the stack gas, if particulate matter adheres to the window, the instrument will no longer read zero when the mirror is put into position. Instead, an upscale reading will be generated by the "dirty" window. The instrument will automatically correct reported opacity values up to a level of 4% opacity. For "dirty window" corrections greater than 4% opacity, an alarm will register, as required by Performance Specification 1.

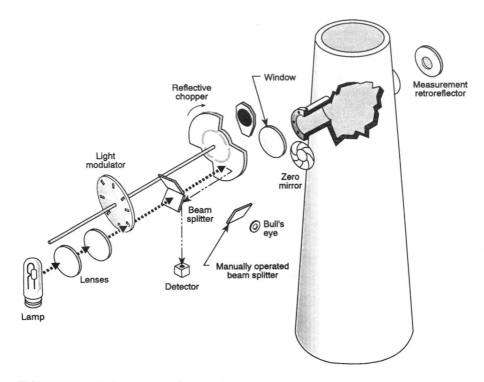

**FIGURE 7-7.**    Reference mode ($I_0$ value) of the Erwin Sick double-pass transmissometer.

One important component of the system is the mechanism used to align the instrument. A manually operated beam splitter is located between the principal beam splitter and the reflective chopper for this purpose. This beam splitter can be moved manually into the path of the projected beam. Light returning from the retroreflector will reflect from the surface of the splitter to a bull's-eye sight on the side of the transceiver. A small circle of light can be observed, which should be centered in the bull's-eye to indicate proper alignment. Adjustments to the instrument mountings are necessary if the alignment is found to be unacceptable.

### The Durag Transmissometer
The Durag Corporation manufactures a transmissometer that is similar to that of the Erwin Sick instrument. Instead of using a rotating chopper to measure the transmittance, a "mode shutter" alternately blocks the light

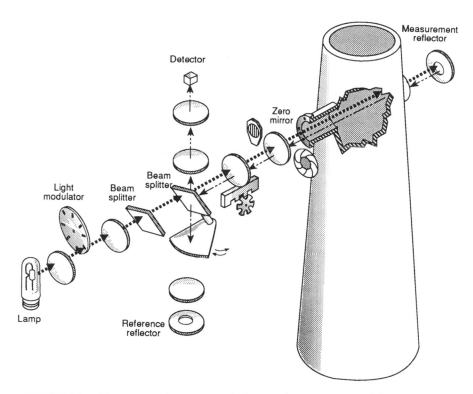

**FIGURE 7-8.**    Durag transmissometer optical system (measurement mode).

beam to make the $I_0$ and $I$ measurements as shown in Figure 7-8. The lamp projects light through a modulator that rotates at a rate of 1200 Hz, to code the transmitted light. The light then passes through two beam splitters. The first beam splitter is used to align the system whereas the second beam splitter is used to split light between a reference reflector and a measurement reflector.

To obtain the transmittance value, $I/I_0$, the mode shutter is positioned to block either the measurement beam or the reference beam. To obtain the measurement value, $I$, the mode shutter is positioned to block the light beam so that it does not impinge on the reference reflector at the bottom of the optical assembly. This allows the part of the projected light that passes through the splitter to proceed through a focusing system and through the flue gas to the retroreflector. The light returns from the retroreflector, strikes the beam splitter, and reflects upward to the detector. In this measurement mode, the detector monitors the light beam for a period of 1 min and 58 s.

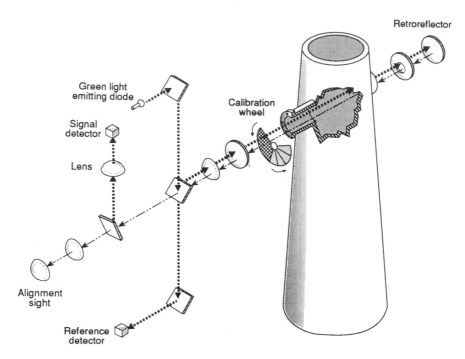

**FIGURE 7-10.**    United Sciences transmissometer optical system (measurement mode).

reading is desired for an EPA-required 24-h zero and span check, both of the sample-and-hold circuits will sample at the same time when the reference zero portion of the chopper is in front of the light beam.

### The United Sciences Transmissometer

The United Sciences instrument also uses a rotating wheel, but incorporates a different optical design (Figure 7-10). Two detectors, one a measurement detector and the other a reference detector, are used to obtain the $I$ and $I_0$ values. The lamp intensity is kept constant using a control loop based on the reference detector signal.

The measurement signal detector is set up to give 0% opacity when no particulate matter is present and 100% opacity when the beam is completely attenuated. As the calibration wheel rotates, the measurement detector senses two different reflectances (Figures 7-11) from the calibration sections of the wheel in addition to the cross-stack attenuated lamp intensity. One of the measured reflectances from the wheel serves as the system "zero" value and the other serves as the upscale calibration value,

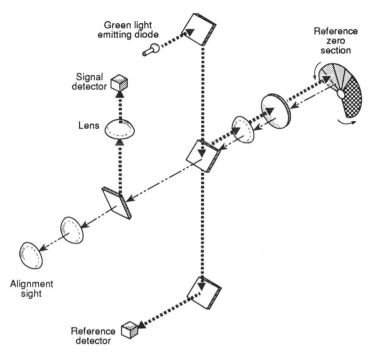

**FIGURE 7-11.**    United Sciences transmissometer optical system: reflection from the reference section.

as in the Thermo Environmental Systems unit. The two reflectances correct for differences between detector responses, electronic and optical changes, and dirt accumulation on the window.

Instead of an incandescent lamp, the United Sciences unit uses a modulated, green-light-emitting diode (LED). Because the LED is electronically modulated, a modulating wheel is not necessary.

### Optical Fiber Techniques

Several instruments have been designed that use fiber-optic cables to guide light to various components located within the transceiver. These systems minimize the use of optical and mechanical components, relying instead on electronic switching and fiber optics to perform the instrument functions. The Lear Siegler/Dynatron transmissometer is the most notable of these units.

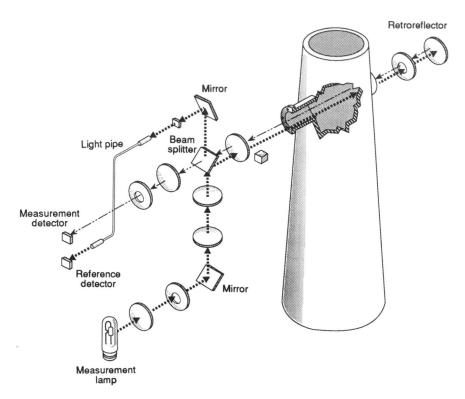

**FIGURE 7-12.**    Lear Siegler/Dynatron transmissometer optical system using a fiber-optic cable: measurement mode.


### *The Lear Siegler / Dynatron Transmissometer*

The Lear Siegler/Dynatron monitor instrument performs the transmittance measurement by using two detectors in a somewhat unique design. Part of the system is shown in Figure 7-12. Tracing the path of the light beam projected from the lamp in the bottom left-hand corner of the figure, the light passes through a condensing lens, through an aperture, to a mirror, and then up to a beam splitter. The splitter separates the light into two beams; one passes through it and bounces off a mirror to enter a fiber-optic light guide. The light travels through the guide to the reference detector to give the $I_0$ signal value.

The measurement beam is projected by reflection from the underside of a beam splitter. It leaves the transceiver and crosses the stack to the retroreflector. The retroreflector returns the light to the beam splitter, where this time it passes through the splitter and is focused on the measurement detector to give the $I$ value.

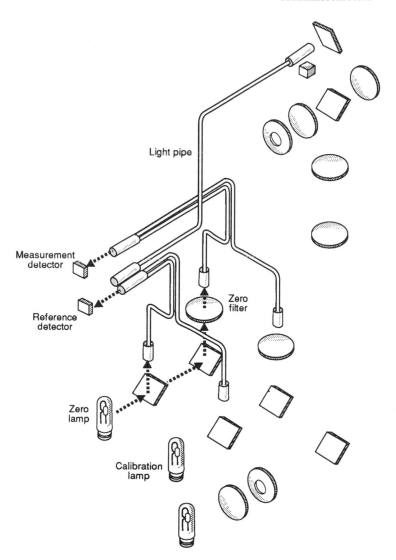

**FIGURE 7-13.**   Lear Siegler/Dynatron transmissometer optical system using a fiber-optic cable: low-level value calibration.

Because two detectors are used, some means must be provided to account for variations in detector sensitivity. This is done by incorporating a multilamp calibration system as shown in Figures 7-13 and 7-14. The calibration system is composed of two parallel units designed to provide a "simulated zero," or low-level, reading and an "upscale," or high-level, reading.

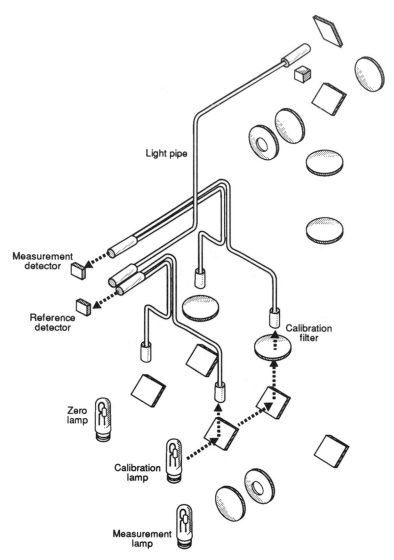

**FIGURE 7-14.**    Lear Siegler/Dynatron transmissometer optical system using a fiber-optic cable: high-level value calibration.

To obtain the low-level reading (Figure 7-13), the measurement lamp at the bottom of the instrument is turned off and the zero lamp is turned on. The light is projected to a beam splitter that divides it. The reflected beam enters a fiber-optic light guide that directs the light to the reference detector. The transmitted portion is reflected from a small mirror to pass through a neutral density filter and enters another light guide. This beam is directed to the measurement detector. This neutral density filter will have a low-scale value (less than 10% opacity).

To obtain the high-level calibration value (Figure 7-14), the zero lamp and the measurement lamp are both turned off. The calibration lamp is turned on and the beam is projected toward a beam splitter, as in the zeroing system. The beam is split; one portion is reflected to pass through a fiber-optic guide to the reference detector. The transmitted portion is reflected by another mirror to pass through a neutral density filter. This filter will have a higher opacity value than the one used in the simulated zero system. After passing through the filter, the light enters another light guide to be directed to the measurement detector. The instrument response is then used to establish the instrument upscale calibration value. This response is essentially the second point of a two-point calibration for the analyzer.

The fiber-optic features of this design are intended to eliminate moving parts in the transceiver. However, an option provided by the company incorporates a moving pseudozero mirror and an upscale attenuator that are similar to the pseudozero mirror operation of the Erwin Sick and Durag systems.

## Other Techniques

Several companies manufacture simple, single-pass units that can be zeroed and calibrated only under clean-stack conditions. These units consist of a light source on one side of the stack and a detector on the other. These monitors may be perfectly adequate for specific applications, but in general will not meet the design criteria of Performance Specification 1 (PS 1). The advantage to using them is that they tend to be lower in cost, tend to weigh less, and are easier to install than the double-pass units.

Single-pass units can, however, be designed to meet PS 1. By using either a zero pipe or fiber-optic cables, calibration systems can be designed that will check the active electronic and optical components of the system. By placing a pipe across the stack and either flushing it with clean air or sealing it to be free of particulate matter, zero values can be

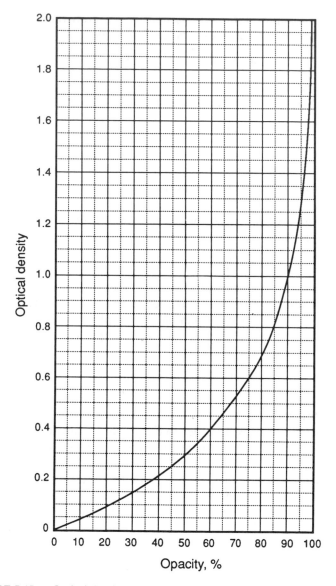

**FIGURE 7-15.**    Optical density versus opacity.

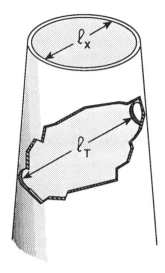

$\ell_x$ = pathlength at emission outlet

$\ell_T$ = pathlength at monitor location

**FIGURE 7-16.**    Stack exit correlation.

where $\ell_T$ = the single-pass path length at the monitor location

By dividing the two expressions, the single-pass optical density at the stack exit can be obtained:

$$D_x = (\ell_x/2\ell_t)D_t \qquad (7\text{-}6)$$

The flue gas is not appreciably compressed at the outlet of a tapered stack. Instead, the velocity increases, so that the particulate matter concentration and the volumetric flow rate will be the same at both locations. Therefore, if $\ell_x$ is less than $\ell_t$, the single-pass opacity will be smaller at the stack exit than at the monitoring location.

Taking the antilogarithm of Equation (7-6) and solving for opacity, the stack-exit corrected opacity value can be obtained:

$$\text{Op}_x = 1 - (1 - \text{Op}_t)^{\ell_x/2\ell_t} \qquad (7\text{-}7)$$

Care must be taken in the use of these equations. In single-pass transmissometers, the term $\ell_t$ stands for the optical path length of the measurement beam. This could be simply the stack or duct diameter at the transmissometer location. For a double-pass system, the light traverses the stack twice, therefore the path length is 2 times $\ell_t$.

**Calibration Filter Selection**

The expressions just given can also be used for selecting transmissometer calibration and audit filters. For example, if an inspector wishes to check the performance of a double-pass transmissometer using an audit filter, the following equation would be used to select the filter (Jahnke 1984):

$$D_{\text{nominal filter value}} = \left( \frac{\ell_t}{\ell_x} \right) D_{\text{desired instrument output value}} \qquad (7\text{-}8)$$

For example, let us say that an inspector needs to purchase a midrange filter that will produce a reading of $D = 0.30$ (50% opacity) when it is inserted into an audit jig. If the monitor path ($\ell_t$) is 5.0 m and the stack-exit diameter ($\ell_x$) is 7.5 m (single-pass values), then

$$D_{\text{nominal filter value}} = \left( \frac{5.0}{7.5} \right) 0.3 = 0.20$$

or

$$Op_{\text{nominal filter value}} = 1 - 10^{-0.2} = 0.37$$

or 37% opacity. A filter having a nominal value of 37% opacity would then be purchased for the audit.

**Combiner Equations**

The optical density expression is also useful in situations where two or more transmissometers are placed in ducts that lead to a single stack such as that shown in Figure 7-17. This arrangement is quite common because it enables the plant operator to monitor the performance of the separate control equipment that exhaust to a common stack. Although the arrangement may be useful for plant control, air pollution agencies often require that the opacity be reported as if a transmissometer were located at the stack exit. Such data can be obtained by taking the readings from the individual transmissometers and using them in a combiner equation. Vendors of transmissometer systems sell optional electronic units that will perform these calculations automatically. Care should be taken in applying these units because a number of assumptions need to be satisfied if accurate values for the combined opacity at the stack exit are to be

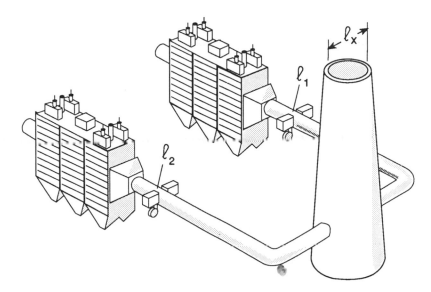

**FIGURE 7-17.**    Two ducts entering a common stack.

obtained. In Figure 7-17, the optical density at the stack exit can be obtained using the following combiner equation:

$$D_x = \frac{D_1 A_1 v_1 (\ell_x/\ell_1) + D_2 A_2 v_2 (\ell_x/\ell_2)}{v_1 A_1 + v_2 A_2} \tag{7-9}$$

where $A_1$ and $A_2$ are the cross-sectional areas of each duct at the point of measurement and $v_1$ and $v_2$ are the flue-gas velocities in each duct. If the duct areas and flue-gas velocities of each duct are identical, this simplifies to

$$D_x = \frac{D_1 (\ell_x/\ell_1) + D_2 (\ell_x/\ell_2)}{2}$$

The opacity at the stack exit can then be obtained from the optical density. The conditions of equal duct areas and gas velocities may not always be met. In such cases, velocity monitors or methods of estimating the duct velocities are required.

## Opacity–Mass Correlations

From the discussion of the Bouguer relationship in Chapter 4, it should be clear that the light scattering and absorbing properties of particles in a flue gas are related to their concentration in the gas. In fact, data obtained from manual particulate traverses [U.S. EPA Reference Method 5 or 17 or ISO Standard 9096 (ISO 1991)] can be correlated with transmissometer data. Standards have been prepared that detail both test procedures and methods for developing these correlations: ISO Standard 10155 (ISO 1992) and VDI (1980).

The correlation can be done easily if the optical density–extinction expression [Equation (4-11)] is considered:

$$D = \frac{A_E c \ell}{2.303}$$

If $A_E$ and $\ell$ are constant, a plot of optical density versus particle concentration should be linear. A common practice in both Europe and the United States is to plot concentration versus extinction. Extinction is defined as $b = D/\ell$, the optical density per unit path length. In terms of the extinction, then, the expression becomes

$$b = A_E c / 2.303 = Kc$$

where $K$ is a constant.

A typical correlation plot is illustrated in Figure 7-18. Note both the confidence and tolerance intervals included in the plot.

The relationship between concentration and extinction depends on the basic assumption that the particle characteristics remain the same as when the correlation was obtained. If the particle size distribution changes, the light scattering behavior may also change. This will cause a spread in the concentration–extinction data. However, the technique may still be used if the spread is not too great (as determined by the width of the confidence and tolerance intervals). Experience has shown the following:

1. Correlations must be made on a unit-by-unit basis, that is, a correlation developed at one coal-fired boiler may not necessarily be the same as that developed at another coal-fired boiler.
2. Scatter in the data will primarily depend on (a) methods used to vary the particulate mass concentration when the correlation is developed and (b) the competence of the source testing team performing the manual sampling.

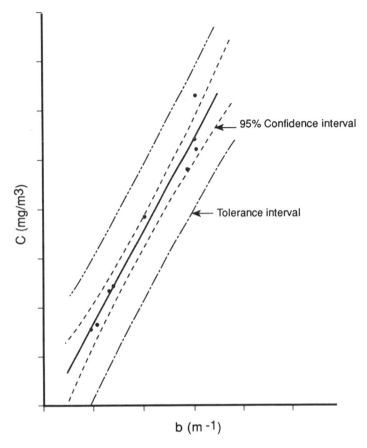

**FIGURE 7-18.**    Plot of concentration versus extinction using a double-pass transmissometer.

The opacity–mass correlation technique is valid only so long as the conditions under which it was developed are representative of the source operation. Changes in operation that lead to significant changes in particle characteristics or the particle size distribution may greatly affect the slope of the correlation line. Changes in fuel, control equipment, or process operation may contribute to this problem. A new correlation should be developed in such situations and such guidance should be provided in any regulatory specification. By checking single points on a routine basis, the continuing validity of the correlation can be assessed.

# 8

# Data-Acquisition and Control Systems—Recording and Reporting CEM System Data

A CEM system is not complete without incorporating a subsystem that records the data produced by the analyzers. The data acquisition system (DAS) provides this record of emissions measurements, both for documenting plant operations and for reporting to the environmental control agency. On a daily basis, the CEM system operator or plant environmental engineer works more with the DAS than with any other CEM subsystem. Activities such as reviewing data, checking calibration values, responding to excess emissions problems, and generating reports are all performed within the DAS. This system's importance cannot be overemphasized.

The actual CEM DAS functions depend heavily on regulatory requirements and the complexity of the monitoring system. The DAS can be as simple as a strip chart recorder, or it can incorporate personal computers (PCs) or plant mainframe computers. The principal function of a CEM DAS computer is to collect and record data. Analyzer analogue signals must be converted to digital signals at some point in order to be processed by a computer; however, many newer analyzers are incorporating their own microprocessors to produce digital outputs that can be processed directly. After the data are recorded, they can be manipulated, converted into different units, averaged, and reported by the computer.

The computer also can provide controlling functions for the analyzers, such as performing automatic daily calibrations and adjustments. Alternatively, these controlling functions can be performed by programmable logic controllers (PLCs), or other devices, so that the computer can be dedicated to recording and reporting data.

188

## CONTROL SYSTEMS

A control system provides for the automatic operations of the CEM system hardware. Functions such as zero and span checks, extractive probe blowback, and alarms for excess emissions and system malfunctions all may be performed automatically. Either programmable logic controllers or the CEM system computer will provide the signals to activate relays or valves for these operations.

Programmable logic controllers have been applied widely in industry to handle the logic, timing, counting, and data transfer in manufacturing operations. PLCs are modular and can be chosen to perform a number of functions, such as triggering automatic functions, providing analog to digital (A/D) signal conversion, registering alarms, and data logging. They can be programmed to perform mathematical calculations, to perform calibration corrections to data, and to average the data.

Data-logger–controllers perform similar functions, but are usually single units designed to receive a fixed number of analogue inputs and to provide a fixed number of outputs. These can be tailored to meet the general monitoring requirements.

An advantage of PLCs and similar control systems is that, using them, the choice of actual data recording and reporting system becomes more flexible. The plant computer or a PC using a standard CEM program can then simply accept inputs without having to be programmed for the control functions of the system.

## RECORDING AND REPORTING DEVICES

There are four devices that can be used to record and report CEM system data, including

1. the strip chart recorder
2. the data logger
3. the PC or dedicated CEM computer
4. the plant mainframe computer

Various combinations of these subsystems are often incorporated into the CEM DAS. Given that data are increasingly recorded, corrected, converted, and stored using computerized systems, a failure in the computer can be catastrophic—worse than an analyzer failure. For this reason, data backup and computer quality assurance are important.

## Strip Chart Recorders

The strip chart recorder is the most basic CEM system recording device and can be an invaluable tool to an instrument operator. Even if a computer system is used to acquire CEM data, a strip chart recorder is useful in providing a backup record. Advantages of the strip chart recorder include the following:

- The continuous, high-speed response of a strip chart recorder is invaluable for system troubleshooting.
- The strip chart recorder provides the most accessible record for the auditor or inspector.
- The strip chart record readily indicates if an instrument is properly "zeroed" and "calibrated."
- Trends in source performance and instrument performance are easily detected.
- The appearance of the strip chart and the annotation of dates, times, strip chart speeds, and so on can indicate how well the instrument is being operated and maintained.

A strip chart recorder should be easy to read and be flexible enough to be used for multiple purposes. A chart width of 6 in. or greater is recommended. The recorder should have a range switch for different inputs, and it should be possible to change the chart speed. Accordion-fold paper that allows for easy folding is convenient for data storage; however, chart rolls can be manually folded for the same effect.

A strip chart recorder should be chosen so that data can be interpolated at a level suitable to the measurement range specified for the CEM system. Changes in the chart trace should be distinguishable at a level better than the drift limit. In a 1000-ppm range, for example, it should be possible to distinguish changes better than 2.5% of the range, which is 25 ppm. Although the strip chart recorders offer many advantages, there are several disadvantages associated with their use:

- Strip chart recorders are high-maintenance devices. Paper and pens must be replaced periodically.
- Recorder failures can result in the total loss of data.
- Multiple recorders can occupy valuable panel space, particularly in the control room.
- Strip chart data can be difficult to store and access.
- In computerized systems, automatic calibration corrections or calculations are sometimes made only on the computer record. Such corrections are performed mathematically by the computer and do not constitute physical adjustments to the calibration potentiometers of the

analyzers. As a result, these changes do not register on the strip chart record, so the strip chart data are difficult to interpret or are virtually useless.

Despite these problems, the strip chart record can be invaluable for "at-a-glance" evaluations of emissions, system troubleshooting, or as back-ups to the computer data. In many plants, strip chart recorders are installed in the control room to monitor opacity, but gaseous emissions data are provided by computer.

## Data Loggers

Data loggers are devices that print data in digital form. They are convenient to use when the amount of effort required to reduce strip chart data to actual numbers becomes excessive and when data manipulation needs are minimal. A data logger is not a sophisticated microprocessor, but it can be designed to perform rudimentary calculations and alarm functions. Data loggers are often used with microprocessors to convert analyzer analogue signals into digital forms that are acceptable to the computer. The data logger can also store the data for backup purposes. For example, battery-backed solid-state, nonvolatile storage cartridges can be used for backup in the case of power failure.

Although data loggers have limited capabilities, they may perform all of the functions necessary for limited applications. However, for full data-reporting capabilities, a computer system may be used in conjunction with a data logger.

## Personal Computers and Dedicated CEM System Computers

A computerized system can process, display, and transmit CEM analyzer data and can support other functions necessary for CEM system operations, such as analyzer control. Various types of systems are available in today's market, supplied either by the CEM system vendor or by companies that develop data acquisition systems. In general, these systems are IBM-compatible, use a 386 or 486 chip, are equipped with hard drives of up to 120 megabytes. Depending on the system, an internal A/D conversion capability may be included. Continuous emissions monitoring data acquisition systems should be designed to minimize the loss of data during power interruption or system failure.

The data acquisition system design offered by a CEM systems vendor often remains fixed. A potential user may ask for special features such as touch screens, laser printers, color displays, and multitasking capabilities, but many of these features come at increased cost and CEM system vendors are not always willing to spend the time and effort to deviate from their existing software.

## The Plant Mainframe Computer

Using the plant mainframe computer for CEM system data acquisition may appear to be an attractive option. If sufficient memory is available, the mainframe generally offers more programming flexibility than a dedicated CEM computer. The ability to incorporate spreadsheet and graphics programs into the CEM software can provide the state-of-the-art displays to which process engineers have become accustomed. The CEM system data will also be directly part of the plant information system and may be more easily transmitted through that system. The main advantage is that if plant personnel are responsible for the system programming, the plant will have control of the software and will be able to modify it as needed, without having to rely on an outside vendor.

However, there are disadvantages with this approach. A plant programmer is generally not dedicated to the CEM system software and other departments make demands on his or her time. Also, the programmer usually does not have experience in the many obscure points associated with CEM regulation and will only learn these over a period of time. It normally takes from six months to a year to develop operating and debugged CEM software; with this in mind, the costs associated with its development may be equivalent to those of a system provided by an outside vendor.

## Telemetry

As an option, the system may be designed to transmit data to the regulatory agency or to corporate offices. The protocol used to transmit data via a dial-up telecommunications system should follow a data telemetry access protocol specified by the requesting organization. Care should be exercised by agencies in requiring real-time CEM system data. Real-time data may not be quality-assured, and knowledge of real-time data may make the agency a party to plant emission problems. Because the agency would have current knowledge of plant emissions, if a plant upset or catastrophic release should occur and no agency action is taken, the public may respond negatively.

## UNITS OF THE STANDARD

Practically every emissions regulation requires that analyzer data be converted into specified units. These units, or "units of the standard," depend upon the regulatory policies of the national or local environmental control agency and the type of source regulated. Mass rate standards are common in Europe, whereas process rate standards are more frequently specified in the United States. Table 8-1 gives several examples of units that are commonly used.

Gaseous emission values may require reporting in one of these forms (transmissometer data is normally reported in percentage of opacity). For example, in the United States, emissions from fossil-fuel-fired steam generator facilities are expressed in pounds per million British thermal units (nanograms per joule); for a discussion of F-factor methods used to perform these calculations see Appendix A and, for example, U.S. EPA (1991b) or Jahnke and Aldina (1979). In the U.S. acid rain program, allowance trades for $SO_2$ emissions are expressed in tons of $SO_2$ per year and in the large furnace orders in the Federal Republic of Germany (FRG 1988), gaseous emission limits are expressed in kilograms per hour. In the United States, for total reduced sulfur (TRS) emissions from Kraft pulp mill lime kilns, emissions are monitored in parts per million by volume, corrected to 10% $O_2$ on a dry basis. How emissions standards are determined affects the design of the CEM system. For this reason, a facility that is purchasing a CEM system should include regulatory requirements in the CEM system request for proposals.

## RECORDING AND REPORTING DATA

Continuous emission monitoring systems can give a wealth of data; however, not all of the data are required to be recorded. These requirements may differ between federal and state governments and certainly between different countries (Bühne 1981). As an example, Table 8-2 gives U.S. federal recording requirements for opacity and gaseous emissions (U.S. EPA 1991a).

These requirements specify the minimum number of data points required to be recorded. Systems are commonly designed to poll the continuous analyzer data for a period of typically every 10 s and then average the numbers obtained over the required period of 6 or 15 min. The 15-min recording period for gas analyzers allows for the installation of analyzer time-sharing systems. In time-sharing systems, one analyzer can be used to measure emissions from two to three stacks (e.g., 5-min sequential measurements for each 15-min period) to meet the specification. The intent of

**TABLE 8-2    New Source Performance Standards Recording Requirements**

*Opacity*    A cycle of sampling and analyzing for each successive 10-s period, and one cycle of data recording for each successive 6-min period

*Gaseous emissions*    One cycle of sampling, analyzing, and data recording for each successive 15-min period

this approach is to reduce the number of analyzers in a CEM system; however, this increases the complexity and decreases the flexibility of the system.

## Data Averaging

Emissions data reporting is further complicated by data-averaging requirements. In most cases, the requirements will necessitate that they be determined by a computerized data-acquisition system for CEM system data.

Block averages and rolling averages are two averages commonly used in regulatory reporting. A block average is an average of sequential data points or an integrated average obtained over a specified time period. Examples of these methods are given next.

### The Block Average

Block averages are used in U.S. federal regulations for the averaging of opacity and gas emissions data. Their application depends upon the source category and the pollutant being regulated; for example, for Subpart D electric utilities in the United States, opacity 6-min block averages are to be calculated from 36 or more data points equally spaced over each 6-min period. For gaseous emissions, 1-h block averages are to be calculated from four or more data points equally spaced over each 1-h period.

Opacity monitoring data are generally recorded and reported in 6-min block averages. Gaseous emissions monitoring data, however, can be recorded and reported as an hourly average, or the 1-h block average can be further averaged over 3 or 24 h before reporting.

### The Rolling Average

Block-averaged data can be further treated by using a "rolling average," which is an arithmetic average of a specified number of contiguous periods. For example, in a 3-h rolling average, three contiguous 1-h averages are averaged (this is shown in Figure 8-1). The 3-h rolling average is used in the U.S. federal reporting requirements for NSPS Subpart D fossil-fuel-fired steam generator facilities.

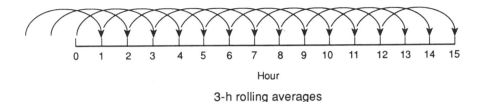

Hour

3-h rolling averages

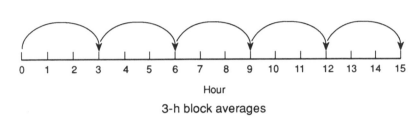

Hour

3-h block averages

**FIGURE 8-1.**    A 2-h rolling average and how it differs from a 3-h block average.

Note from the figure that at the end of 4 h, the 1-h value of the previous 3-h average is dropped, and the 4-h value is averaged with the remaining 2 h. Similarly, the average "rolls" along from hour to hour. This differs from a block average, given that in a 24-h period, there will actually be 24 contiguous 3-h periods. If one were to average blocks of 3-h periods, only $\frac{24}{3} = 8$ averaged values would be obtained.

A 30-day rolling average is specified in the reporting requirements for U.S. NSPS Subpart Da (electric utility steam generating units). This is an average of 30 contiguous 24-h averages, as shown in Figure 8-2. This average is similar to the 3-h rolling average, except that 30 days of daily data are averaged. In a 30-day period, 30 averages (not one average) would be obtained, after the initial 30-day lapse. To allow for CEM system shutdown time, the rolling average may not actually be the average of 30 days of data, and the daily data may not be composed of an average of 24 h of data. [*Note*: The specification allows a minimum of 18 h of hourly data (block averaged) to be included in each 24-h block average. It also allows a minimum of 22 daily averages to be used in calculating a 30-day rolling average. This works out to a CEM system availability of approximately 75%; that is, the CEM system is available to give data 75% of the time. Because CEM system availability for gas monitoring systems is currently between 90 and 97% (McCoy 1990), the specification is particularly generous.]

emission data are used for control operations and environmental reporting, a computerized system is usually necessary. Facilities often increase
the sophistication of the DAS and reporting system when it becomes
apparent that too much time is spent by senior personnel in reducing the
emission report data.

## SUGGESTED FUNCTIONS FOR CEM
## SYSTEM SOFTWARE

In the preceding discussion, general requirements were given for recording
and reporting emissions data. It is the responsibility of the CEM system
programmer to incorporate these requirements into a consistent set of
programs. Programming can be either rudimentary or sophisticated; however, users are becoming more demanding because of their experience
with the flexibility and graphics capabilities of commercial software. User
concern should not be directed so much at the CEM system hardware
because hardware is continuously developing and basically is a set of tools
with which the programmer works. Instead, the primary concern of the
user should be with what comes out of the system and how easy it is to
work with. Some initial suggestions for the user in this regard are given
here.

Software should be written in a standard computer language (e.g.,
BASIC, PASCAL, or FORTRAN) and should consist of an optimum
combination of a real-time operating system and program to provide
(1) the data-base structure, (2) an easy-to-use operator interface, (3)
user-selectable parameters, (4) report formats, (5) editing commands,
and (6) archiving features. Specifically, the computer system should do the
following:

### 1. *General Features*

- The system should acquire data from the emission monitors as it is
  generated, while simultaneously allowing for operator input and report
  generation.
- The data base should incorporate the entire past quarter's data and the
  data generated during the current quarter.
- The user should be able to edit and demand reports for any period in
  the previous quarter, as well as for the current quarter.
- Status and editing codes should be incorporated into the reporting
  system for conditions such as out-of-service instruments, abnormal cali-

brations, insufficient samples, data substitutions, and edited data invalidations. The user also should be able to enter additional codes after system installation.

- The user should be able to configure the system on-site to change parameters such as emission limits, data correction constants, alarm set points, calibration tolerances, and range scales. Also, textual information fault codes should be able to be defined.
- Printed reports should be generated from the data base on demand. The capability to insert comments from the keyboard into the report also should be available.
- Editing provisions should be made for conditions such as reason code entry or modification, comment entry, data invalidation (due to CEM malfunction), plant malfunction and emergency conditions, and nonoperating conditions.
- Provision should be made for the entry of manual test or alternate monitor data in case of CEM system malfunction.
- Menu-driven programs, help screens, or other user aids should be incorporated into the software, to minimize user training time with the system.

2. *Calculations.* The system should correct pollutant data to the units of the standard. Parameters such as combustion efficiency, $SO_2$ removal efficiency, or emissions in units other than the standard (e.g., kilograms per hour, pounds per hour, nanograms per joule, or parts per million corrected to 6% $O_2$) may also be calculated.

3. *Screen displays.* The system should provide a continually updated display of parameters measured and corrected by the monitoring system. The displays should provide easily readable formats in one screen or several screens.

4. *Recording.* Data should be recorded on 3.25- or 5.25-in. disks so that they are readable by IBM PC–compatible disk drives.

5. *Reports.* Reports should be able to be generated by the DAS on command. The system should be capable of producing hard copy of reports such as the following:

- *CEM system daily report.* A daily CEM system status report that includes monitor calibration data, hourly averages, excess emission data, and monitor availability
- *Quarterly emissions reports.* A report prepared for each compliance parameter for each unit of the facility (i.e., opacity, $SO_2$ concentration)

**6.** *Inputs*. The CEM data-acquisition–control system should provide inputs to the plant mainframe computer to display desired CEM data for the control room operator.

Although commercial data-acquisition systems may not offer all of the features listed here, the capabilities are available in today's technology to meet such requirements. It is expected that the systems will evolve as users begin to require increased programming sophistication.

### References

Bühne, K. W. 1981. Automatische Verfahren zur Auswertung von kontinuierlichen Emissionsmessungen. *Staub-Reinhalt. Luft* 41:175–183.

Federal Republic of Germany (FRG)—Federal Ministry for the Environment, Nature Conservation and Nuclear Safety. 1988. *Air Pollution Control Manual of Continuous Emission Monitoring*. Bonn, Germany.

Jahnke, J. A., and Aldina, G. J. 1979. *Continuous Air Pollution Source Monitoring Systems—Handbook*. EPA 625/6-79-005.

McCoy, P. G. 1990. The CEM nation: An analysis of U.S. EPA's database—1988. In *Proceedings—Continuous Emission Monitoring: Present and Future Applications*, Air and Waste Management Association, Pittsburgh, pp. 10–36.

McRanie, R. D. 1990. Continuous emissions monitoring: Looking beyond the horizon. *Power*. 12:45–48.

Paley, L. R. 1984. *Technical Guidance on the Review and Use of Excess Emission Reports*. EPA 340/1-84-015.

U.S. Environmental Protection Agency (EPA). 1990. Amendments to standards of performance for new stationary sources—reporting requirements. *Federal Register*, 55 FR 51378 (December 13, 1990).

U.S. EPA. 1991a. Monitoring requirements. *Code of Federal Regulations*, 40 CFR 60.13. Superintendent of Documents, Washington, DC.

U.S. EPA. 1991b. Method 19. Determination of sulfur dioxide removal efficiency and particulate matter, sulfur dioxide, and nitrogen oxides emission rates. *Code of Federal Regulations*, 40 CFR 60 Appendix A. Superintendent of Documents, Washington, DC.

### Bibliography

Jaye, F., Steiner, J., and Larkin, R. 1978. *Resource Manual for Implementing the NSPS Continuous Monitoring Regulations. Manual 3—Procedures for Agency Evaluation of Continuous Monitor Data and Excess Emission Reports*. EPA 340/1-78-005c.

Nazzaro, J. C. 1985. Continuous emission monitoring system approval, auditing and data processing in the Commonwealth of Pennsylvania. In Transactions—

Continuous Emission Monitoring: Advances and Issues (J. A. Jahnke, Ed.). Air Pollution Control Association, Pittsburgh, pp. 175–186.

Quinn, G. C. 1978. Recording instruments—A special report. Part 1. *Power* 12:s1–s28.

Quinn, G. C. 1979. Recording instruments—A special report. Part 2. *Power* 1:s9–s18.

Ruger, D. W. 1985. A data acquisition system for quarterly reporting of compliance data. In *Transactions—Continuous Emission Monitoring: Advances and Issues* (J. A. Jahnke, Ed.). Air Pollution Control Association, Pittsburgh, pp. 440–451.

Ruger, D., and Ketchum, R. 1981. Automatic reporting of continuous emission monitoring system data. In *Proceedings—Specialty Conference on Continuous Emission Monitoring: Design, Operation and Experience.* Air Pollution Control Association, Pittsburgh, pp. 253–267.

Siedhoff, T. E. 1979. Design of an Automated Emission Monitoring and Reporting System. Paper presented at the Air Pollution Control Association Meeting, Cincinnati. Paper 79-35.3.

Webb, R. O., Shell, M. A., Kesterson, G. M., and Lanter, T. A. 1985. Design of a computer-based CEM and air quality data acquisition system for Alabama Power Company. In *Transactions—Continuous Emission Monitoring: Advances and Issues.* (J. A. Jahnke, Ed.). Air Pollution Control Association, Pittsburgh, pp. 52–63.

# 9

## Certifying CEM Systems

Procedures for certifying CEM systems have been established by the U.S. EPA, the International Standards Organization (ISO), and a number of European countries. National certification procedures are established as regulatory requirements and must be followed if the CEM system is to provide data for the control agency. Different approaches can be taken to CEM certification. However, most approaches involve a comparison to an independent reference method, either manual or automated. Table 9-1 gives the U.S. EPA performance specification procedures that have been established. The U.S. performance specifications for NSPS sources are given in 40 CFR 60 Appendix B.

It has been found that a monitor may work well at one facility but that the same model may give erratic data at another. It has also been found that installations differ. Therefore, it is not the monitor that is the controlling element but the total system (probe or stack interface, conditioning system, analyzer, and the controller–data-acquisition system) that must be evaluated to prove that accurate data can be provided. Certification must be done on a case-by-case basis to account for stack conditions and the unique characteristics of the installed CEM system.

It is important to remember that these test procedures have been established to certify the CEM system as installed and that a certified system should generate data that are representative of the stack emissions. The installation specifications, performance specifications, and test procedures were all developed with this intent. However, it should be noted that the performance specifications do not evaluate the continuing operation of the system. The long-term operation of the system depends on the quality assurance (QA) program developed for the system.

TABLE 9-1    Performance Specifications

| Number | Type of System | FR Reference | Date Promulgated or Revised |
|--------|---------------|--------------|------------------------------|
| PS 1 | Opacity | 48 FR 13322 | 3/30/83 (revised) |
| PS 2 | $SO_2$, $NO_x$ | 48 FR 23608 | 5/25/83 (revised) |
| PS 3 | $O_2$, $CO_2$ | 48 FR 23608 | 5/25/83 (revised) |
| PS 4 | CO | 50 FR 32984 | 8/5/85 (promulgated) |
| PS 5 | TRS | 48 FR 32984 | 7/20/83 (promulgated) |
| PS 6 | Velocity, mass emission rate | 53 FR 07514 | 8/9/88 (promulgated) |
| PS 7 | $H_2S$ | 54 FR 18308 | 10/2/90 (promulgated) |

## INSTALLATION SPECIFICATIONS

Two overriding principles for CEM system installation are (1) that measurements must be representative of stack emissions and (2) the sampling location should be accessible for system maintenance and repairs. It is often difficult to satisfy both criteria, particularly in existing sources where sampling considerations were not addressed during duct or stack design. Also, in newly constructed sources, sampling criteria are often secondary in importance to the practical concerns of design engineers.

### Accessibility

Because the frequency and quality of maintenance are directly related to the accessibility of a CEM system, the location of a probe or monitor is an important consideration in any CEM system design. Sheltered locations that can be reached by elevators or stairways are preferred to those that are not sheltered and can only be reached by ladder. An installation in the boiler house or in the annulus between the stack and stack liner provides shelter at the measuring point. An exposed location on a roof or off a catwalk is less preferable, but is often chosen so that measurements can be made at a representative location. In some cases, a CEM systems vendor will install a small, modular shelter suitable for one person to perform maintenance at the measuring point. For stack installations of in-situ analyzers, factors such as vibration, lightning, ambient atmosphere, and temperature extremes, in addition to accessibility and representativeness, must be considered for long-term operation.

### Representativeness

Although accessibility is an important consideration in identifying a site for monitoring systems, if a location does not offer the possibility of obtaining a representative sample of the flue gases, little will be gained. In a regulatory sense, "representative" means that "the pollutant concentration or emission rate measurements are directly representative or can be corrected so as to be representative of the total emissions from the affected facility..." (U.S. EPA 1991b). Representativeness for transmissometers means that the monitor should be placed at a location where the measured opacity corresponds to the actual opacity of the emitted flue gas. When considering the overall accuracy of the CEM system in terms of stack emissions, representativeness of the sample being measured is as dependent on the measuring location as it is on the accuracy of the analyzers performing the measurements.

### Gas Stratification

Flowing gases are generally well-mixed, but stratification can occur when there are differing temperatures or when dissimilar gas streams intersect (see, for example, Figure 6-4). Air leaking into a duct, the combining of two streams of process gas into a single stack, or the reintroduction of scrubber bypass gas into a flue all can result in such stratification. In combustion sources, air inleakage usually occurs near the preheaters. Columns of gas with high unmixed $NO_x$ concentrations have even been observed from burners. The problem is further complicated because this stratification is not only spatial—stratification can also change temporally (i.e., as a function of time). As process load or other conditions change, the gas distributions can vary dynamically.

### Particle Stratification

Particles may not be uniformly distributed in a flue gas. Because of their momentum, particles may not be able to change direction as rapidly as the flue gas flowing through the plant exhaust system. This will cause the development of gradients in particle number concentrations at certain locations in the ductwork (Figure 9-1). The particles may even become stratified in layers over relatively long sections of a flue. Particle stratification depends largely on four factors:

- the size of the particles
- the flue-gas velocity
- the distance between the point of interest and a disturbance
- the nature of the disturbance

The disturbance can be a bend in the ductwork, air inleakage, or a disturbance in the actual stack exit.

In long, horizontal ducts, gravitational settling can lead to the stratification of larger particles toward the bottom. The problem of stratification

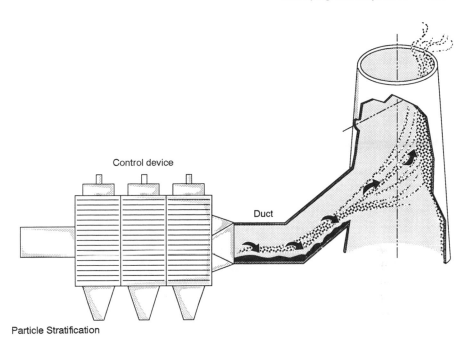

Control device

Duct

Particle Stratification

**FIGURE 9-1.**    Particle stratification in ducts and stacks.

can be compounded by low flue-gas velocities because the carrying capacity of the gas for large particles will be reduced. Bends, fans, and obstructions will also affect the distribution of particles in these layers. Also, a helical flow pattern can be produced from the tangential entry of gas from one flue into another (Figure 9-2). The swirling pattern of gas flow may make it difficult to find a transmissometer path that will produce an average opacity representative of the actual emissions.

Another problem occurs when the location being sought is to be representative in terms of mass concentration instead of opacity. Manual source testing methods for particulate matter (U.S. EPA 1991a; ISO 1991) have their own criteria for sampling locations, which may differ from those for transmissometers. A transmissometer measures opacity, not mass concentration. Some of the best locations for measuring opacity would be very difficult locations from which to extract a manual reference method sample. (An area of high turbulence is an example of such a location.) In practice, the important factor is that particles of all sizes should be well-mixed in the stack gas. Areas of turbulence (where there is sufficient time to provide mixing) are good monitor locations, but may be poor for manual sampling because of reverse and fluctuating gas flows.

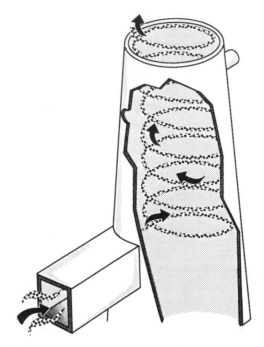

**FIGURE 9-2.**    Helical flow patterns due to tangential entry.

## Quantifying the Degree of Stratification

The degree of stratification in a duct or stack can be quantified. One method of quantification has been proposed (U.S. EPA 1979) that involves traversing the stack or duct and obtaining gas concentration values. An example of a scenario for a rectangular duct would be to sample at nine sampling points of a balanced matrix. The degree of stratification at each sampling point can be calculated as

$$\text{percentage of stratification at point } i = \frac{(c_i - c_{\text{ave}})}{c_{\text{ave}}} \times 100$$

where $c_i$ = concentration of the pollutant at point $i$

$c_{\text{ave}}$ = average of the nine concentrations

The sampling plane is said to be stratified if any value is greater than 10%.

Another approach to quantifying stratification is to calculate the mean value and standard deviation from the data obtained in the stratification test (Brooks and Williams 1976). Using the standard deviation as a measure of spread, one can use a value of $2\sigma$ (where $\sigma$ is the standard deviation) to define the stratification from the mean concentration value in the sampling plane. For example, if $2\sigma = 11\%$, then the probability is 95% that a point measurement taken anywhere in the sample plane will deviate from the mean concentration by less than $\pm 11\%$. This treatment, however, assume that the individual measurements are normally distributed.

When performing a stratification test, it is good practice to sample at a single point over the entire sampling period (e.g., Elam and Ferguson 1985). This is easily done using an instrumental technique. The data obtained can be used to determine if gas concentrations are changing as a function of time as well as spatially. If the concentration varies at the point over the sampling period, the traverse data will be difficult to interpret.

Stratification tests are difficult to perform well and are costly if a complete characterization of pollutant flow distributions is needed. Also, many CEM systems are installed in new plants and must be on-line at the time of plant start-up. Because sampling locations are decided upon during plant design and construction, it is virtually impossible to conduct stratification tests to guide CEM installation decisions in new plants.

## EPA Recommended Locations

As a means to facilitate the installation of CEM systems at representative locations, guidance has been provided (U.S. EPA 1991d). Some suggestions for measurement location are listed here.

1. *Probe locations.* The location should be at a point at least two equivalent duct diameters downstream from (1) the nearest control device, (2) the point of pollutant generation, or (3) a point at which a change in the pollutant concentration or emission rate may occur. The location should also be at least one-half an equivalent diameter upstream from the exhaust or one-half an equivalent diameter upstream from a control device when measuring uncontrolled emissions. However, these criteria do not guarantee that flue gas concentrations are not stratified at such locations.

2. *Measurement point in the duct or stack.* After choosing the measuring location for sampling probes or monitors, it is further suggested that (1) the probe tip be greater than 1.0 m from the stack or duct wall and that (2) the probe tip monitors either within or over the centroidal area of the stack or duct cross section. For path in-situ monitors, the measurement path is recommended (1) to exclude the area bounded by

a line 1.0 m from the stack or duct wall and (2) to have at least 70% of the path within the inner 50% of the stack or duct cross-sectional area to pass through the centroidal area.

Diluent monitors should sample at the same point as the pollutant monitor. If the $O_2$ or $CO_2$ concentrations vary between the two points because of air inleakage effects, the CEM system may not certify as a system.

If the CEM system does not certify and if it is determined that this failure is due to the monitor location, then the system may have to be relocated. It may be possible to apply a correction technique to avoid relocation, but this is often difficult. For situations of highly stratified flow, there is little evidence that there is a relationship between the composition at a single point and the average concentration when the spatial distribution of pollutant concentrations changes with time.

### Transmissometer Installation Guidelines

The transmissometer installation specifications established by the U.S. EPA consider two aspects: the location and the measurement path at the location. The actual opacity at locations meeting the EPA criteria may or may not be representative of the actual emissions; however, the criteria establish a compromise between having to perform extensive testing to prove representativeness and the arbitrary selection of convenient monitoring locations (U.S. EPA 1991c).

The criteria for measurement locations are as follows:

1. The monitor is to be placed in a location where the opacity measurements are representative of the total emissions. Locations where the stack gas is well-mixed are considered primarily in choosing sites.
2. The monitor is to be accessible to permit routine maintenance, such as window cleaning and blower maintenance. Accessibility is also important for the performance of calibration audits and alignment checks.
3. The monitor is to be located downstream from all particulate control equipment.
4. Water droplets (condensed water vapor) are not to be present at the monitoring location.
5. If the monitor responds to ambient light, it is to be located at a point where ambient light is not present (away from either the top of the stack or where light leaks into the ductwork).

A consideration for selecting the measuring path is that if particle gradients exist in the flue gas, the measurement path should pass through

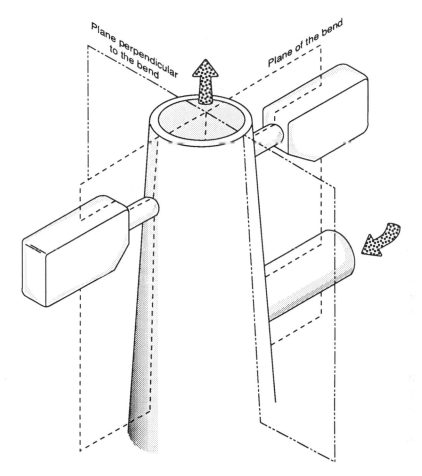

**FIGURE 9-3.**    Plane of the bend.

those gradients. If particulate matter is forced to the outside wall after a bend, the path should be in the plane of the bend, not perpendicular to it (Figure 9-3). In this example, a measurement path in a plane that is perpendicular to the bend may pass through a region of lower particle concentration if the particles are stratified. Paths through the plane formed by the bend would be more representative of the total emissions in this situation.

To reach compromises between representative measurements and accessibility, a number of measurement path criteria have been established (U.S. EPA 1991c). These criteria define monitor paths in terms of how far they are located from horizontal and vertical bends in a duct. The distances are again expressed by the number of duct diameters that the site is located from a bend. Because the installation requirements and path criteria are based partly on scientific principles and partly on regulatory compromises, it is best that plant and agency representatives agree on the monitoring site before installation.

## U.S. EPA PERFORMANCE AND EQUIPMENT SPECIFICATIONS

The primary U.S. EPA performance specifications for gas monitors are found in 40 CFR 60 Appendix B (U.S. EPA 1991b). Similar specifications are found in U.S. federal programs that address issues such as acid rain (40 CFR 75) and hazardous waste incineration (40 CFR 266). In the U.S. performance specifications, there are two major criteria that CEM systems must meet: (1) calibration drift and (2) relative accuracy, which are defined as follows.

**Calibration drift** "The difference in the CEM system output readings from the established reference value after a stated period of operation during which no unscheduled maintenance, repair or adjustment took place."

**Relative accuracy** "The absolute mean difference between the gas concentration or emission rate determined by the CEM system and the value determined by the reference method's (RM) plus the 2.5% error confidence coefficient of a series of tests, divided by the mean of the reference method tests or the applicable emission limit." That is,

$$RA = \frac{|\bar{d}| + |CC|}{\overline{RM}} 100$$

where $|\bar{d}|$ = the mean difference between the reference method result and CEM result

$|CC|$ = the confidence coefficient

$\overline{RM}$ = the average of the reference method values obtained in the test series

The calibration drift specification is not difficult for most commercial CEM systems to satisfy. In some cases where an analyzer may drift beyond the specification, automatic correction made by a microprocessor may bring the system within the specification. The actual specifications that are to be met are given in Table 9-2.

TABLE 9-2    Performance Specifications for Gas Measuring CEM Systems

| PS | Gases | Calibration Drift | Relative Accuracy |
|---|---|---|---|
| 2 | $SO_2$, $NO_x$ | 2.5% of span | 20% of RM value in units of the standard<br>10% of applicable standard (for stds. > 130 ng/J)<br>15% of applicable standard (for stds. > 86 and<br>  < 130 ng/J)<br>20% of applicable standard (for stds. < 86 ng/J) |
| 3 | $O_2$, $CO_2$ | 0.5% | 20% of RM value or 1.0% (whichever is greater) |
| 4 | CO | 5% of span<br>for 6 of 7<br>test days | 10% of RM mean value in units of the standard<br>5% of applicable emission standard<br>  (whichever is greater) |
| 5 | TRS | 5% of span<br>for 6 of 7<br>test days | 20% of RM mean value in units of the standard<br>10% of the applicable standard<br>  (whichever is greater) |
| 6 | Flow | 3% of span | 20% of RM mean value in units of the standard<br>10% of the applicable standard<br>  (whichever is greater) |
| 7 | $H_2S$ | 5% of span<br>for 6 of 7<br>test days | 20% of RM mean value in units of the standard<br>10% of the applicable standard<br>  (whichever is greater) |

The accuracy of a measured value is an expression of its relationship to a standard or true value. In the case of source emission measurements, the true value is not determinable, so accuracy is generally expressed with respect to a reference method value. Reference method values are obtained either by using manual wet chemical techniques or by alternate, instrumented methods.

The U.S. EPA procedure for calculating CEM accuracy, uses a method for the comparison of data pairs and results in the expression of "relative accuracy" (Natrella 1963). The relative accuracy is composed of two terms, one expressing the average deviation of the CEM value from the reference value, the other giving an estimate of the "spread" or "precision."

The relative accuracy expression essentially gives an estimate of accuracy for only one point, the average value of the emissions (RM) at the time the relative accuracy test was performed. It has little statistical relevance at other emission levels that the CEM system might measure. A modified relative accuracy expression that substitutes the value of the emissions standard in place of RM, is used to accommodate problems associated with the expression at low emission values. Such modifications, and others, are regulatory constructs and have little statistical meaning. It also should be noted that the relative accuracy is dependent upon the accuracy and precision of the reference method itself.

Despite these shortcomings, the relative accuracy expression has served as a useful criterion for accepting or rejecting newly installed CEM systems. The tendency has also been to use relative accuracy as a measure of the total performance of a CEM system.

## PERFORMANCE SPECIFICATION TEST PROCEDURES

The performance specification test (PST) is the test required by the U.S. EPA to accept or certify CEM systems. The test evaluates how well the system performs under the physical and environmental conditions at the plant. The ability to sample the flue gas, response to line voltage fluctuations, sensitivity to duct vibration, ambient temperature, and ambient atmosphere are implicitly checked in the calibration drift test or the relative accuracy test. The PST is not a laboratory evaluation of an instrument under controlled conditions, but a realistic evaluation under plant operating conditions.

For competitive purposes, a CEM system vendor must often guarantee that the installed system will meet the specifications and, consequently, the PST determines success or failure for the CEM system vendor. Companies that purchase the systems often demand such guarantees in their contract with the vendor and may hold partial payment or apply other penalties until the system is certified. However, even with such guarantees and inducements, a system may not pass all the test criteria the first time. Most often, problems can be resolved by tracking down mistakes in the system design or by modifying the system to meet the special demands of the plant. In some cases, particularly with the application of new technology, the entire monitoring concept may prove unworkable and the system will have to be replaced by one that is more suitable. Therefore, these tests are considered extremely important and must be carried out correctly: The calculations must be performed properly, and a well-documented report of the test must be prepared. The following section describes how each of these should be done.

### Conducting the Test

The PST is usually conducted by a source-testing contractor, who commonly sends out a mobile van with instrumentation capable of performing automated reference methods (e.g., U.S. EPA Methods 3a, 6C, 7E). Two or more source-testing personnel generally conduct the relative accuracy tests—the number of personnel depending on the complexity of the CEM system. The calibration drift test can be carried out easily by plant personnel prior to conducting the comparative tests used to determine the relative accuracy.

Representatives from the CEM system vendor and the regulatory agency will often be present to observe and watch for any problems. Through the development of a test plan prior to the tests, all parties involved should determine and understand their roles during the week of testing. The agency representative is particularly important because problems often occur during plant start-ups that may require modification of procedures. In such cases, the ability of the agency to approve or disapprove such changes is central to completing the test.

A pretest meeting should be held before the date of the PST; this will help minimize the lack of coordination and other problems that frequently occur. The meeting should include the following personnel: the plant operator or supervisor; the plant environmental engineer or CEM coordinator; a control agency representative; a source test contractor representative; and a CEM system vendor representative. Topics at the meeting should include the following:

1. status of plant operation at test time
2. status of plant emission control equipment at test time
3. current operational status of CEM system
4. test schedule (for calibration drift and relative accuracy for each unit and parameter of the CEM system)
5. calibration gases and test procedures to be used
6. exceptions
7. test report requirements

There should be a tour of the plant to locations where the reference samples will be taken and where the data will be obtained (control room, CEM room, or other). All questions concerning sampling or monitoring locations should be resolved before the test team arrives.

Unfortunately, pretest meetings are frequently not held, either because the expense is considered too great or the affected parties are too busy. The miscommunication and misunderstandings that result can lead to the rejection of the test data and a repeat of the test.

### The Calibration Drift Test

In the 40 CFR 60 Appendix B Performance Specifications, the zero- and high-level drift tests examine the CEM system's ability to hold its calibration over a period of time. The calibration drift test is conducted over a period of 168 h when the plant is operating at more than 50% of normal load. The CEM system is then evaluated at 24-h intervals for 7 consecutive days.

The calibration drift test is conducted by introducing calibration gases into the CEM system. An in-situ or nonextractive monitor may determine the zero- and high-level calibration drift by producing a mechanical instrument zero and checking the calibration with a gas cell or optical filter. These calibration procedures should first be checked with the regulatory agency.

Under 40 CFR 60 Appendix B, for a CEM system that accepts calibration gases, the test is performed in the manner shown:

1. Day 1—Zero gas (or one with a low-level value, 0–20% of the span value) is introduced into the system. The system is zeroed.
2. Day 1—High-level value calibration gas is (50–100% of span) introduced into the system. The system is calibrated to this value.
3. Day 2—Zero gas is injected into the system after 24 h. The CEM system value is recorded. (If desired, the system may be adjusted at this point to the zero value set on day 1.)
4. Day 2—After injecting zero gas, the high-level value calibration gas is injected. The CEM system reading is recorded. (If desired, the system may be adjusted at this point to the value set on day 1.) If the system was not reset to zero prior to injecting the high-level gas, then the reading obtained in step 3 is subtracted and the result recorded. (*Note:* If periodic automatic or manual adjustments take place to bring the system into calibration, the calibration drift must be determined immediately before these adjustments take place. In some CEM systems, the system microprocessor will automatically zero and calibrate the system daily. Obviously, if the calibration drift test is conducted after such a procedure, there will not be any observable drift.)
5. Days 3–8—Steps 3 and 4 are repeated.

*Calibration Drift Calculations*
Calibration drift is calculated by percentage, using the units of reference gas, cell, or optical filter and dividing by the span value:

$$d_{CD} = \frac{(\text{cylinder gas reference value}) - (\text{monitor value})}{(\text{span value})} 100 \quad (9\text{-}1)$$

where $d$ = the difference between the data pair

*The Relative Accuracy Test*
The relative accuracy test is the most important part of the performance specifications and the most expensive to perform. As emphasized previously, coordination between all parties involved in the test is particularly important (Lindenberg 1981).

The relative accuracy test is conducted to determine if a CEM system will give data (within the specified limits) that can be compared with data obtained using the compliance test methods (the U.S. EPA reference methods of 40 CFR 60 Appendix A). Unlike calibration drift, a relative accuracy determination may not be based solely on data from individual analyzers, but may include the data from pollutant, diluent, or flow monitors to perform the calculation in terms of "units of the standard." For example, $SO_2$ emissions standards, expressed in terms of ng/J or lbs/$10^6$ Btu for fossil, fuel, fired steam generators, require an $F$-factor calculation using $SO_2$ and $O_2$ (or $CO_2$) data. As another example, $SO_2$ emissions standards for municipal waste combustors are corrected to 7% of $O_2$ on a dry basis that can incorporate data from $SO_2$, $O_2$, and $H_2O$ (or equivalent) monitors.

### Reference Method Location and Traverse Points
Relative accuracy is relative only to the reference method determinations and does not necessarily indicate the accuracy of the CEM data with respect to the true value of the stack emissions. To obtain a "truer" value of the stack emissions, the reference method testing is performed on a three-point traverse rather than at a single point in the stack or duct.

The reference method *sampling points* are different than those specified for the CEM system. The sample points are chosen to provide representative samples over the duct or stack cross section. As a minimum, samples are taken on a three-point traverse on a measurement line that passes

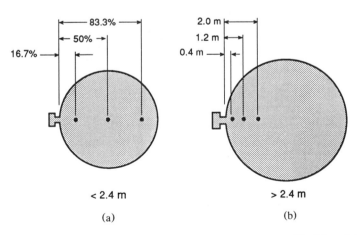

**FIGURE 9-4.**    Reference method traverse points on a measurement line (a)  $< 2.4$ m and (b)  $> 2.4$ m (minimum requirements).

through the centroid of the stack or duct and in the direction of any expected stratification. For a measurement line less than 2.4 m, samples are taken at points that are 16.7, 50, and 83.3% of the line (Figure 9-4a). For larger ducts or stacks with a measuring line greater than 2.4 m and where stratification is not expected, sampling points are specified at 0.4, 1.2, and 2.0 m (Figure 9-4b). (This second option is not allowed after wet scrubbers or where two gas streams with different pollutant composition combine.) Samples are to be taken within 3 cm of these traverse points.

The reference method *sampling locations* are the same as those specified for the CEM system, which are at least two equivalent diameters downstream from a control device and one-half an equivalent diameter upstream from the effluent exhaust or control device.

The reference method sampling point should not interfere with the CEM system probe. A distance of 30 cm or 5% of the equivalent diameter of the cross section (whichever is less) is specified as an appropriate separation.

### Relative Accuracy Test Procedures

The principal sampling strategy for the relative accuracy test is to take CEM readings and reference method samples at the same time. This may seem obvious, but it is sometimes forgotten or not corrected for in the test program. Care must be taken to account for the response time of the CEM system relative to that of the reference method (Figure 9-5).

Figure 9-5 illustrates that the length of the umbilical cord to the CEM system may be important by introducing a lag time before the analyzer sees the sample. If the reference method instantaneously samples or if its response is faster, the time difference must be considered when comparing the data. Note from the figure that the CEM system reading is obtained at a time ($\Delta t$) from when the reference method reading is taken.

A problem also arises when grab samples are taken by a manual reference method or when a time-shared CEM system samples over only a few minutes to perform one "cycle of operation for each 15-min period." This often results in an attempt to compare data obtained over a few seconds or a few minutes with a sample or with sample data averaged over a 21-min-period. This may be satisfactory for a steady process, such as a utility boiler, but for a highly variable process, such as municipal waste combustion, this comparison may not be suitable to evaluate CEM system performance. Technically, in the case of grab samples, the arithmetic average of the CEM system value recorded at the time of each grab sample should be used instead of the average value over the test run of 21 min.

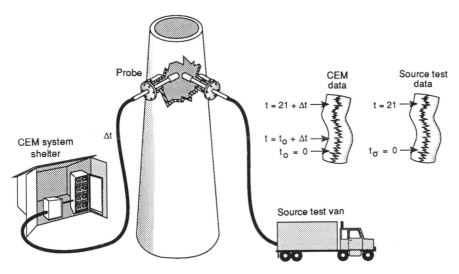

**FIGURE 9-5.**    Correlation of reference method data and CEM data in response time.

Such problems in comparing data can often be overcome by using automated reference methods (U.S. EPA 1991d). Strip chart data or microprocessor-generated data can be compared and evaluated more readily than data obtained from the manual reference methods. Data obtained from the reference methods and the CEM system must all be corrected to the same basis, typically to standard conditions of temperature and pressure [20° C and 1 atmosphere (atm)] dry, and corrected for dilution using the $\%O_2$ or $\%CO_2$ data.

Because there are many areas in data interpretation that may be ambiguous, it is imperative that the control agency, the source owner or operator, and the source tester agree prior to the test how the data will be obtained, compared, and calculated (e.g., Jorgenson and Highberger 1985).

### Obtaining the Test Data

Reference method data is obtained at each of the three points on the traverse line when conducting the test. The following test options are available:

1. *Manual reference method (integrated sample).* Sample 7 min at each point (Figure 9-6a).
2. *Alternate reference method (integrated sample).* Sample 7 min at each point (instrumental methods) (Figure 9-6a).

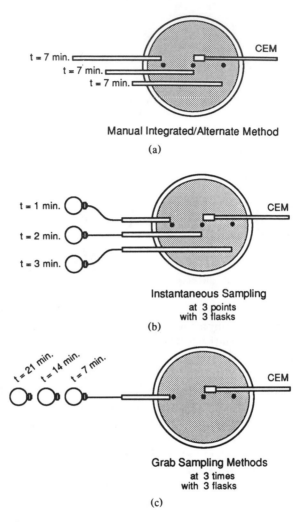

**FIGURE 9-6.**   Procedures for obtaining the reference method samples.

3. *Manual reference method (grab samples).* Sample the three points simultaneously (within 3 min) (Figure 9-6b) or sample at equal intervals over a period of 21 min or less (Figure 9-6c).

   These procedures are repeated to give a total of nine data sets. Each set of data is obtained within a period of 30–60 min, so an efficient test team should be able to conduct the relative accuracy test in one day. The tester may conduct 12 runs and collect 12 sets of data. The tester may reject, at his or her discretion, three sets of data. No statistical basis is necessary for rejecting any set of data, but all of the data, including those that have been rejected, must be reported.

Diluent and moisture measurements should be obtained simultaneously with the pollutant measurements, although PS 2 gives the option that they can be obtained during the 30- to 60-min sampling period. The sampling error associated with these measurements will be as significant in the relative accuracy calculations as those associated with the pollutant measurements; therefore, it is best to be as consistent in diluent and moisture measurements as possible.

## RELATIVE ACCURACY TEST CALCULATIONS

The relative accuracy of the CEM system is determined by comparing the instrument-measured emission values (in units of the standard) to the reference method analysis. As previously noted, the reference method may either be a manual method or an alternate instrumental method. When comparing CEM system data with manual reference method data, a few more calculations are necessary than if the alternate methods are used. In a manual reference method, data are reported from the analytical laboratory in terms of milligrams of pollutant per dry standard cubic meter (dscm) or dry standard cubic feet (dscf). These results are corrected to a standard pressure of 760 mmHg (29.92 in. Hg) and a standard temperature of 293 K (528 R). All data for the reference methods and the CEM system must be given on a consistent, dry basis. Because in-situ analyzers, dilution probes, and some extractive systems (those that analyze hot gas) measure on a wet basis, a moisture correction may be necessary.

The relative accuracy is determined by dividing the absolute value of the mean difference between the reference method data and the CEM system data plus a confidence coefficient |CC| by either the average reference method value or applicable standard as follows:

relative accuracy

$$= \frac{|\text{arithmetic mean of differences}| + |\text{confidence coefficient}|}{\text{mean of reference method values (or applicable standard)}} 100$$

(9-2)

independent as the actual relative accuracy check.

### The Test Report

After the tests have been successfully completed, the test report is prepared. The test report documents the tests performed on the CEM system and presents the test results. The report should be well-organized, readable, and complete. All data, including raw data necessary to recalculate any of the reportable parameters, should be included in the report.

Performance specification 2 includes a minimum list of items that are to be reported:

1. results of calibration drift and relative accuracy tests
2. data sheets
3. calculations
4. strip charts or computer printouts of CEM system response
5. cylinder-gas or calibration-cell concentrations and certifications

Some agencies have additional report requirements or require special formats and tables.

## INTERNATIONAL STANDARDS ORGANIZATION STANDARDS FOR GAS MONITOR CERTIFICATION

The International Standards Organization (ISO) has prepared standards for $SO_2$ and $NO_x$ CEM systems. The $SO_2$ standard has been published as ISO Standard 7935 (ISO 1990). The $NO_x$ standard is currently in draft status as ISO DIS 10849. As in the U.S. CEM performance specifications, the ISO standards reflect the philosophy that each CEM system, *as installed*, must meet a set of minimum performance criteria.

### General Format of the ISO $SO_2$ and $NO_x$ CEM Standards

The format of the international $SO_2$ and $NO_x$ CEM standards differ from U.S. EPA Performance Specification 2 (40 CFR 60 Appendix B) by incorporating (1) required standards and test procedures and (2) recommended standards and test procedures. The ISO standards also provide a discussion of CEM monitoring techniques and the physical–chemical principles of measurement of monitoring instrumentation. Because the standards are intended for use by all countries, some of which may not have an

**TABLE 9-4    ISO-Required and Implicit Performance Characteristics for $SO_2$ and $NO_x$ CEM Systems**

| Performance Characteristic | Numerical Value |
|---|---|
| ISO-Required Performance Characteristics | |
| Lower detection limit | $\leq 2\%$ of span |
| Interference rejection | $\leq \pm 2\%$ of span for $SO_2$ |
| | $\leq \pm 4\%$ of span for $NO_x$ |
| Response time | 200 s |
| Integral performance, $S_A$ | $\leq \pm 2.5\%$ of span for $SO_2$ |
| (standard deviation of | $\leq \pm 5\%$ of span for $NO_x$ |
| reference method test data) | |
| Implicit Performance Characteristics in Test Procedures | |
| Synergistic interference effects | $\leq \pm 20\%$ between measured and |
| | calculated interference |
| Analyzer linearity | $\leq \pm 2\%$ from calibration curve |
| Bias check | $\|d\| - \|cc\| \leq 2\%$ of span |

existing technical base in environmental monitoring, it has been felt that descriptive information of this nature is useful.

### The Performance Characteristics

The main performance characteristics (analogous to U.S. "performance specifications") of the ISO CEM standards are given in Table 9-4.

The ISO recommended performance characteristics are not required to be met, but the view is that by meeting them the system will have a better chance of meeting the requirements given in Table 9-4. These additional performance characteristics are given in Table 9-5. Note that the zero drift and span drift specifications are only recommended in the ISO standards,

**TABLE 9-5    ISO-Recommended Performance Characteristics**

| Performance Characteristic | Numerical Value |
|---|---|
| Zero drift | $\leq \pm 2\%$ of span for $SO_2$ and $NO_x$ |
| Span drift | $\leq \pm 4\%$ of span for $SO_2$ and $NO_x$ |
| Temperature-responsive zero drift | $\leq \pm 2\%$ of span for $SO_2$ and $NO_x$ |
| Temperature-responsive span drift | $\leq \pm 3\%$ of span for $SO_2$ |
| | $\leq \pm 4\%$ of span for $NO_x$ |

in contrast to U.S. EPA Appendix B and Part 75 specifications that require that calibration error and drift be determined.

### ISO Test Procedures

The ISO performance characteristic (specification) test procedures exhibit two basic difference from the U.S. methods:

1. Several of the tests may be conducted in the laboratory.
2. The reference method testing is conducted over the period of "un-attended operation," typically seven days.

Reference method testing may be conducted by using either manual reference methods or an automated method that employs an analyzer that uses a measurement principle different than that of the installed analyzer.

### ISO Laboratory (or Installed System Test Procedures)

Tests that may be performed either in the laboratory or on the CEM system installed at the site are conducted as follows:

**(a)   *Lower Detection Limit (Required Specification)***
The test is to be conducted on the complete CEM system as set up in the laboratory or at the site. The procedure involves injecting zero gas at the probe, or as close as possible to the probe, a minimum of 30 times, to obtain the readings. For in-situ path monitors, flow-through gas cells may be used. The test is to be done as rapidly as possible to avoid effects of zero drift or temperature effects.

The lower detection limit is given by the following expression:

$$x = x_0 + 2s_{x0} \qquad (9\text{-}7)$$

where $x_0$ = the mean of the readings
$s_{x0}$ = the standard deviation of the readings

This is basically a test of the noise level of the system at the zero value and any offsets at zero that are particular to the system.

**(b)   *Interference Rejection (Required Specification)***
The interference rejection test is designed to determine the effects of other flue-gas constituents on the $SO_2$ or $NO_x$ measurements. The test is performed by injecting into the CEM system possible interfering gases at concentration values corresponding to their typical levels in the flue gas.

The gas is to be injected, again, at the probe of the complete CEM system set up either in the laboratory or on-site.

The interference is to be determined for each individual interfering gas. In addition, a mixture of all of the interfering gases must be made up and injected into the system. The interference parameter $S$ is determined in the following manner:

Given the known value $p_{si}$ of the $i$th interfering gas in the test cylinder, obtain the response of the CEM system, $x_{si}$, and obtain the ratio (*note:* calculations are performed on a mass concentration basis, milligrams per cubic meter, not in parts per million)

$$S_i = \frac{x_{si}}{p_{si}} \tag{9-8}$$

where $p_{si}$ = a known mass concentration (in milligrams per cubic meter) of the interfering gas

$x_{si}$ = the value measured on the CEM system (expressed in milligrams per cubic meter) in response to injecting the gas

$s_i$ = the contribution (in milligrams per cubic meter) of interfering gas to the $SO_2$ or $NO_x$ reading, per unit concentration (in milligrams per cubic meter) of interfering gas

If the ratio $s_i$ is not linear with respect to the change of $p_{si}$, determination of the nonlinear dependence of the CEM system on the interfering gas is required.

The total interference, $S$, is calculated for the sum of the interfering gases, using the $s_i$ values as follows:

$$S = \frac{100}{\text{(span value)}} \sum_{i=1}^{n} s_i p_{Mi} \tag{9-9}$$

where       $S$ = the calculated value for total interference effect of a mixture of interfering gases

(span value) = the full scale range of the analyzer

$s_i$   is defined as before

$p_{Mi}$ = the mass concentration (in milligrams per cubic meter) of the $i$th gas in the mixture

The interference is limited to 2% for $SO_2$ analyzers and 4% for $NO_x$ analyzers, for the sum of interfering effects due to maximum concentrations of all possible interfering gases.

(c)  *Synergistic Interference Effects (Required—Given Implicitly in Test Procedures)*

Synergistic effects between the possible interferant gases are also required to be determined. For a combustion source, this mixture may be composed of $CO_2$, CO, $SO_2$, $NO_x$, and $H_2O$. Each gas will have a known mass concentration $p_{Mi}$ in the mixture. The value $R$, corresponding to the response of the CEM system to the mixture, is measured by injecting the mixture.

The calculated value of $S$ is then compared with the measured value $R$. If the two values agree within 20%, the effect of the combination can be neglected and the interference effects for other mixtures of these compounds can be calculated. If it is known that there are no synergistic interference effects between the gases, the individual interference effects $s_i$ need only be calculated.

Interference effects can usually be noted in the system relative accuracy tests; however, an understanding of interferences can be helpful in purchasing a system and in determining if they caused a failed accuracy test.

(d)  *Response Time (Required Specification)*

CEM system response time is determined by injecting calibration gas at the probe of the complete system either installed at the site or assembled in the laboratory. The calibration gas is to have a value between 50 and 90% of full scale.

Response time is the average time interval between injecting the gas and reaching 90% of the concentration value. The response time is required to be less than 200 seconds for the *system* (in contrast to the U.S. EPA specifications of 15 min).

(e)  *Analyzer Linearity (Required Specification—Given Implicitly in Test Procedures)*

The CEM system analyzers are to be checked for linearity. This may be done either in the laboratory or on the system as installed. Because nonlinearity is viewed by ISO as an effect due only to the analyzer and its principle of detection, the checks may be made by injecting calibration gases directly at the analyzer and not through the total CEM system.

In this test, five uniformly distributed gas concentrations (e.g., 20, 40, 60, 80, and 90% of full scale) are injected into the analyzer. A gas divider (dilution) device may be used in this test, rather than using individual cylinders for each concentration.

The analyzer is viewed as being linear if the deviation of a measured value from the linear relation is not greater than $\pm 2\%$.

If the relationship is found to be nonlinear, it is recommended that 10 points be used to establish a curve and that the curve then be used in determining the emission values. This procedure does not constitute a pass/fail test for the analyzer, but establishes a check on the analyzer linearity or nonlinearity. For either situation, the relationship determined may be programmed into the system.

### (f)  Temperature-Responsive Zero and Span Instability (Recommended Specification)

This test can only be conducted in the laboratory, by using a temperature-controlled chamber. The complete CEM system is placed in a climate chamber in which the temperature can be varied from 5 to 35° C at 10 K intervals, or the permissible ambient temperature range given by the manufacturer of the equipment. The analyzer is initially zeroed and calibrated at the monitor calibration port. At temperatures of 5, 15, 25, and 35° C, zero and span gases are injected and the system readings are noted.

For the temperature-responsive zero drift, the difference of the readings from one temperature to the next higher or lower temperature point are calculated relative to full scale.

For the temperature-responsive span drift, the differences of the readings from one temperature to the next higher or lower temperature point are calculated relative to the value of the calibration gas.

The monitor is acceptable if the zero drift is less than ±2% for a change to each temperature level and if the span drift is less than ±3% or ±4% for a change to each temperature level. The test itself may be difficult to perform if the total CEM system is to be checked for this specification. In the United States, typically only the analyzers are checked for temperature drift effects.

### (g)  Zero and Span Drift (Recommended Specification)

Zero and span drift may be determined either in the laboratory on the total CEM system or the CEM system as installed at the site. This specification differs significantly from U.S. specifications in that Performance Specification 2 and the Part 75 rules require the test to be conducted only on the system as installed at the site. U.S. EPA rules are not inconsistent with the specification, but are more stringent.

The test is to be conducted over the period of "unattended operation," typically seven days. During that period, zero and span gases are to be injected at the probe into the CEM system. No adjustments or corrections to the CEM system are to be made during this period (note that it is a period of "unattended operation").

For the zero readings, the differences of the readings between the first day and the last day relative to the span value (full scale) constitutes the zero drift.

For the span readings, the differences of the readings between the first day and the last day relative to the calibration (span) gas value constitutes the span drift.

## ISO Field Test Procedures

(a)  *The Integral Performance Characteristic (Required Specification)*
The only ISO performance characteristic that is required to be determined on the CEM system installed at the site is the "integral performance characteristic," $s_A$, the standard deviation of the differences between data pairs (CEM–reference method). The ISO integral performance test (the ISO equivalent of the U.S. relative accuracy test) is performed over the period of unattended operation (typically seven days). During this period, a minimum of 10 reference method data sets are to be taken, although 30 are recommended. The test is not conducted in one day, but over the seven-day period.

Either manual reference methods or automated methods may be used in conducting the test. The automated methods, however, must use a measurement technique different than that of the installed CEM system. Before each run, the zero and calibration are checked at the analyzer and also at the probe. CEM system measurements are compared to the reference method values over the time period in which the reference measurements were taken.

The integral performance characteristic $s_A$ is calculated from the following expressions.

$$s_A = \sqrt{s_D^2 - s_M^2} \tag{9-10}$$

where $s_M$ = the precision of the reference method expressed as a standard deviation; must be determined separately
$s_D$ = the standard deviation of the difference between data pairs, $d_i$, as given in Equation (9-6)

The CEM system meets the specification if $s_D$ is less than 2.5% of span (range) for $SO_2$ CEM systems or less than 5.0% of span for $NO_x$ CEM systems, that is,

$$\frac{s_A}{(\text{span})} \times 100 \leq \begin{cases} \pm 2.5\% & \text{for } SO_2 \text{ CEM systems} \\ \pm 5.0\% & \text{for } NO_x \text{ CEM systems} \end{cases}$$

Calculations are to be performed in units of milligrams per cubic meter. The calculation does not address accuracy criteria in terms of units of the standard (e.g., nanograms per joule or kilograms per hour) or any corrections for dilution performed with the use of $O_2$ or $CO_2$ analyzers.

Note that the integral performance characteristic $s_A$ is referenced with respect to the span value of the analyzer and not the reference method value, as is done in the U.S. Appendix B performance specifications. This method corresponds with the practice of instrumentation engineers, who refer to matters of precision with respect to the measuring range. This is practical in the sense that the type of analyzer one builds for measuring at high levels (concentrations) may be quite different from one required to measure at low levels. For example, a precision of $\pm 5\%$ for an analyzer with a range of 0–1000 ppm gives some uncertainty over $\pm 50$ ppm. If one wished to make precise measurements in range of 0–100 ppm with this analyzer, there would clearly be a problem. On the other hand, a precision of $\pm 5\%$ on an analyzer designed to measure over a range of 0–100 ppm gives an uncertainty of $\pm 5$ ppm. The trick, of course, is to design an analyzer that can measure at such a level of precision on the 0–100 range.

In contrast, the U.S. EPA performance specifications are expressed as a percentage of the reference method value, as we have seen earlier. Because the average pollutant concentration should lie within 40–70% of span, U.S. specifications are implicitly related to the span value. This formalism is somewhat confusing to an instrument engineer because it is first necessary to know the average value of the emissions from the emission source before the span can be determined. Although the U.S. "relative accuracy" only has meaning at the "average" reference method value, this may not be a problem in a regulatory sense because, by definition, this is where the source should be operating. However, instrument engineers are not accustomed to thinking in this manner because the precision becomes a moving target, depending on the reference method value. An instrument engineer, instead, prefers to establish instrument characteristics with respect to fixed decades or preestablished ranges.

### (b)   The ISO Bias Check (Required Specification—Given Implicitly in Test Procedures)

The mean value of the difference between the data pairs is calculated to determine the presence of systematic error during the seven-day performance test. A bias, or systematic error, is considered to be present if

$$|d| \geq 2\frac{s_D}{\sqrt{n}} \tag{9-11}$$

If

$$\frac{|d| - 2s_D/\sqrt{n}}{(\text{span value})} 100 \geq 2\% \qquad (9\text{-}12)$$

the monitor has failed the bias test, and the cause of the bias must be determined and eliminated.

The expression $2s_D/\sqrt{n}$ is essentially equivalent to the confidence coefficient CC of the U.S. EPA Appendix B, $CC = t(s_D/\sqrt{n})$, where $t$ is the $t$ value of the Student's $t$-test, having a value of 2.262 for 9 data points and a value of 2.093 for 30 data points.

The ISO bias test allows a certain amount of bias in a CEM system because it generally is difficult to remove all sources of systematic error. However, bias greater than 2% of the span value is not allowed, and the cause of this error must be found and eliminated from the CEM system. Possible causes of bias are errors in cylinder gas values, temperature and pressure effects, interferences, and so on. A similar bias test has also been incorporated in the U.S. EPA Part 75 CEM rules.

## TRANSMISSOMETER PERFORMANCE SPECIFICATIONS

Performance specifications for transmissometers differ from those for gas monitoring systems. These are found in Performance Specification 1 of 40 CFR 60 Appendix B. The ISO has no comparable specifications. Gas monitoring systems can be readily checked using manual or instrumental reference test methods. However, as discussed in Chapter 7, there is no scientifically independent or reasonably convenient method of checking the accuracy of stack-mounted transmissometers. Design specifications, therefore, have been established to guide the development of opacity-monitoring instruments that hopefully will provide data comparable to that obtained by a visible emissions (VE) observer. As mentioned in Chapter 7, each transmissometer can be tested to verify that it meets these specifications, or a certificate of conformance to the specifications can be obtained from the vendor.

Before a transmissometer is installed, a number of preliminary adjustments and tests must be performed. These include conducting a calibration error test, conducting a response time test, and adjusting the simulated zero to the true zero when the transmissometer is set up in a clean environment (U.S. EPA 1991b).

Another set of tests is conducted after the monitor is installed. The instrument is aligned; then it is shown that the instrument can operate on

**TABLE 9-6    Performance Specifications for Transmissometer Test Procedures**

| Parameter | Specifications |
|---|---|
| Laboratory tests | |
| Calibration error[a] | ≤ 3% Opacity |
| Response time | ≤ 10 s |
| Field tests | |
| Conditioning period[b] | ≤ 168 h |
| Operational test period[b] | ≤ 168 h |
| Zero drift (24 h)[a] | ≤ 2% opacity |
| Calibration drift (24 h)[a] | ≤ 2% opacity |
| Data recorder resolution | ≤ 0.5% opacity |

[a]Expressed as the sum of the absolute value of the mean and the absolute value of the confidence coefficient.

[b]During the conditioning and operational test periods, the CEM system must not require any corrective maintenance, repair, replacement, or adjustment other than that clearly specified as routine and required in the operation and maintenance manuals.

the stack or duct. Values for zero drift and upscale drift are then determined as part of the test procedures. The performance specifications that are to be met are listed in Table 9-6.

### Preliminary Tests and Adjustments

The preliminary tests may be conducted by the monitor vendor after manufacture; however, the intent of Performance Specification 1 is that the test are to be conducted at the source, using the CEM data acquisition system to provide the test data. If the vendor performs the tests, certified test results will be needed for proper documentation of the performance specification report submitted to the agency.

For these tests, the monitoring system is set up in a laboratory, control room, or any other reasonably clean and convenient environment at the plant (Figure 9-7). The transceiver assembly and retroflector assembly are separated by a distance equivalent to the distance between their points of attachment on the stack or duct. If possible, this distance should be determined by direct measurement rather than from engineering drawings. The flange lengths are added to obtain the flange-to-flange distance.

Next, the instrument is adjusted so that it will produce an output that is correlated to the stack-exit opacity. In practice the operator merely needs to calculate the ratio $\ell_x/\ell_t$ (or $\ell_x/2\ell_t$), depending on the instrument) and to set the value using switches or potentiometers located in the remote control unit. Remember that $\ell_t$ is the length over which the flue

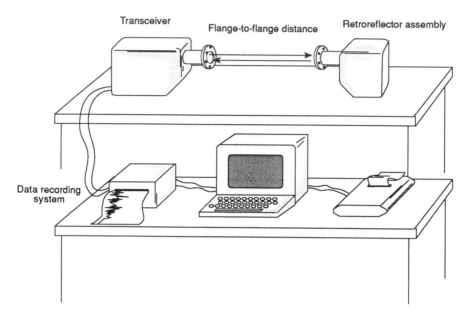

**FIGURE 9-7.**    Example laboratory arrangement for transmissometer preliminary tests.

gas is measured. As such, it is bounded by the walls of the stack or is the length of the slot of a slotted tube. It is not the flange-to-flange distance.

The instrument is then turned on and aligned using the alignment sight. The retroreflector (or receiver unit) is adjusted until the maximum response is reached by the system. Because the monitor is now measuring only laboratory air, it should read zero opacity. At this time, the simulated zero system of the transceiver assembly (for double-pass monitors) is adjusted so that it corresponds to the measurement of zero over the laboratory path.

In systems with zero mirrors, this requires that the iris on the zero mirror be adjusted. Other systems require electronic adjustments. After the zero adjustments are made, the internal upscale calibration value is determined. This usually involves actuating the calibration cycle of the instrument and noting the instrument signal produced when the upscale filter or reflector is in the path of the light beam. The value of this filter must be in the range that is required for the particular installation.

### Calibration Error Test
Next, the calibration error test is conducted. The calibration error test is performed by placing three different neutral density filters (calibration

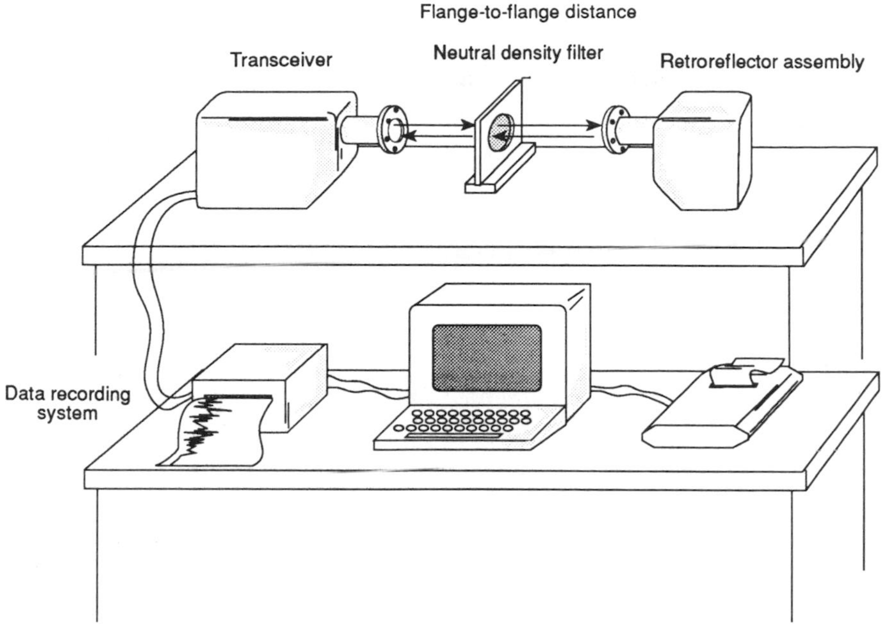

Flange-to-flange distance

Transceiver    Neutral density filter    Retroreflector assembly

Data recording system

**FIGURE 9-8.**    Transmissometer calibration error test.

attenuators) in the light path of the monitor. The laboratory setup allows the filters to be replaced at the midpoint between the transceiver (transmitter) and retroflector (receiver) assemblies (Figure 9-8).

The test requires that the filters be placed at this point rather than in the instrument housing. This provides a better check on the instrument collimation and simulates a measurement made at a point in the path where particulate matter will be present after the monitor is installed. The filters used in this test are required to have values lying within certain ranges. They must also be calibrated against a laboratory spectrophotometer or be certified that they were calibrated by the vendor or an independent laboratory. Three filters are specified: a low-, a mid-, and high-range filter. The filter values correspond to opacity values as observed at the stack exit. They are not necessarily the opacity value at the monitor location, so they must take into account the stack-exit correlation factor.

The actual calibration error test is very simple. First, the low-range attenuator is placed midway between the transceiver (transmitter) and retroflector (receiver), to obtain a monitor reading. Next, the mid-range filter is placed in the path, then the high-range filter. The measured opacity is recorded on a data sheet, and the procedure is repeated until five separate measurements are obtained for each filter. The measurements for each filter are to be made nonconsecutively. When placing the attenuator in the beam, care should be taken to avoid light reflections. Light could possibly reflect from the filter surface to the detector, giving high-opacity readings. This can be minimized by inserting the filters into the beam at a slight angle.

After the 15 measurements are made, the calibration error is calculated. The calibration error is required to be less than 3% opacity for each filter.

### Response Time Test

The response time is the amount of time it takes the opacity monitoring system to display, on the data recorder, 95% of a step change in opacity. The response time test used to determine this value is as simple as the calibration error test. All that is required is the high-range attenuator and a stop watch. The attenuator is placed in the path of the beam, and the time it takes for the instrument to go from its zero value to 95% of the attenuator value is measured and recorded. This is the upscale response time. After the attenuator value has been reached, the attenuator is removed. The time it takes for the monitoring system to go from the attenuator value down to 5% of the value is measured and recorded. This is the downscale response time. This procedure is repeated until five upscale times and five downscale times have been logged. The 10 response time values are then averaged together to obtain the average response time in seconds. The average response time is required to be less than 10 s.

## Installation

After the laboratory tests have been completed, the monitor is installed on the duct or stack. By this time, ports, flanges, access platforms, electrical power, and so on should all be in place at the site. The blower connections are made; the blowers are turned on; then the transceiver and retroreflector assemblies are mounted. Preliminary alignments and adjustments are made and the instrument is turned on so that the light beam can cross the stack. The system can now be optically aligned.

An optical alignment sight or telescope is used to help adjust the transceiver and retroreflector assemblies. When the appropriate light pattern is observed on the sight, one can assume that the system is optically aligned. Ideally, this alignment procedure is to be done when the facility is not operating and no particulate matter is present in the optical path. If no particulate matter is present, the simulated zero can be checked against the true, across-the-stack zero as described in the discussion on laboratory tests.

After the facility is started up, the alignment is checked again. Heated stack walls may cause some misalignment, or vibration may cause problems that were not present when the plant was not operating. If the instrument becomes misaligned under these conditions, it should be realigned. (Under such conditions, the simulated zero should still correspond to the true cross-stack zero.)

If it is not possible to install the monitoring systems when the facility is down, all that can be done is to assume that the simulated-zero–true-zero adjustments made in the laboratory will remain the same after the instrument is mounted on the stack. However, at the first opportunity the alignment and zero adjustments are required to be verified. This may be during a scheduled outage period for maintenance or it may be at a time when malfunctions force the facility to be out of service.

## Transmissometer Field Tests

### Conditioning Period
The next step in the field test procedures shows if the transmissometer system works after it has been aligned and zeroed. This means that the monitor must operate without requiring unscheduled maintenance or repairs for a period of at least one week. The monitor must analyze the flue gas for opacity and permanently record the data obtained during the week. This period is called the conditioning period. Zero and upscale calibration checks are to be conducted once each day, and the optical alignment is to be checked at the end of the period. If the compensation for dirty windows reaches 4% during this period, the windows may be cleaned. The required conditioning period serves several purposes. It gives the instrument operator the chance to become familiar with the system, and it provides a shakedown period for the system components. If the monitor fails before the seven-day period (168 h) is finished, the conditioning period must be repeated after the system is repaired. It is also intended that the source operate during the entire 168 h. If the facility breaks down or runs intermittently, the times and dates of shutdown are to be recorded. The clock is stopped during this time, but is started again after the source resumes operation.

### Operational Test Period
The last part of the PST is the operational test period. With the instrument operating properly, all that has to be done is to obtain the normal

24-h zero and upscale calibration readings and to perform a few calculations. The plant environmental engineer or instrument operator usually can perform these calculations.

During the 168 h of the operational test period, data are taken so that the zero drift and upscale calibration drift can be calculated. The operational test period docs not have to follow immediately after the conditioning period but, again, the plant must be operating and the instrument must be monitoring the flue-gas opacity during the 168 h.

The test is carried out by recording, on the first day of the 168-h period, the simulated zero reading and the upscale attenuation reading. Twenty-four hours later, the readings are taken again. This is continued for at least seven days until seven simulated zero readings and seven upscale attenuation readings are obtained. Zero adjustments and upscale calibration adjustments may be made, but only after each 24-h period. The data must be taken before these adjustments are made and the new zero and calibration readings are taken as the initial readings for the next 24-h period. Similarly, for window cleaning, data must be taken before cleaning, not after. If no adjustments are made, the readings obtained at the end of the 24-h period constitute the initial readings for the next 24-h period.

A special requirement for the operational test period is that if either the simulated zero or the upscale attenuator readings drift by more than $\pm 2\%$ during a 24-h period or cumulatively, adjustments must be made. Windows must be cleaned or the instrument must be electronically zeroed and calibrated.

Performance Specification 1 requires that readings uncompensated for drift must be obtained. This is a very important point. The instrument recorder or computer output may be given only in terms of compensated values. In many transmissometers, the determination of uncompensated readings requires someone to set a switch manually. In other instruments, the recorded output is given only as compensated output; however, the amount of compensation usually can be determined by some manual procedure. Obtaining the data for the PST during the operational test period, therefore, will require day-to-day attention.

The upscale calibration drift test is conducted along with the zero drift test. After the zero reading is obtained at the end of a 24-h period, the upscale attenuator is moved in place with the instrument before a reading of the upscale value is taken. This is an acceptable practice because in the calculations the zero drift is subtracted from the calibration drift before the data are averaged. The upscale reading at the end of a 24-h period provides the initial reading for the next 24-h period. If the calibration is reset after an upscale reading is recorded, the value set is then the initial reading for the next period.

The only real effort needed during the operational test period is for the person conducting the test to make sure that uncompensated zero readings are recorded. The operational test is not so difficult that it requires the services of a contractor or a representative from the instrument manufacturer. The agency may send an observer to see if the test procedures are being properly performed, but normally such observers will be present only for a day or two. If, however, gaseous emissions monitors are being evaluated during this same period, many more people may be involved, and they may be present for a longer period of time. The manual source tests required as part of the performance specifications for gaseous emissions monitors necessitate more planning and more quality control procedures than are necessary for evaluating the acceptability of opacity monitors.

## Calculations

There are three basic equations used in calculating performance values for transmissometer systems.

1. The arithmetic mean, $\bar{x}$

$$\bar{x} = \frac{1}{n} \sum_{i=1}^{n} x_i \qquad (9\text{-}13)$$

where $x_i$ = the difference between a data pair
a. Calibration error:

$$x_{\text{calibration error}} = \text{opacity}_{\text{meter}} - \text{opacity}_{\text{filter \#1}}$$

b. Zero drift:

$$x_{\substack{\text{zero drift} \\ \text{(24 h later)}}} = \text{opacity}_{\substack{\text{zero reading} \\ \text{(Initial reading)}}} - \text{opacity}_{\text{zero reading}}$$

c. Calibration drift:

$$x_{\text{upscale drift}} = \text{opacity}_{\substack{\text{upscale reading} \\ \text{(24 h later)}}} - \text{opacity}_{\substack{\text{upscale} \\ \text{(initial reading)}}}$$

where $n$ = the number of data pairs

It is very important to keep the plus or minus sign when performing the summation.

2. The confidence coefficient CC is expressed as previously [Equation (9-5)].
3. The error Er as defined in PS 1 for the performance parameters is

$$\text{Er} = |\bar{x}| + |\text{CC}| \qquad (9\text{-}14)$$

where $|\bar{x}|$ = the absolute value of the data pairs (this is important because the mean could be negative as well as positive)

$|\text{CC}|$ = the absolute value of the confidence coefficient (positive)

Note that the equation for error (Er) differs from that given in PS 2 and PS 3. In these specifications, $|\bar{x}| + |\text{CC}|$ is divided by the averaged reference method values (or the value of the standard). Examples of calculations using these expressions are given in Jahnke (1984).

## The Test Report

The purpose of the transmissometer test report is to document the tests performed on the purchased instrument and to present the test results. The report should be well-organized, readable, and complete. All data necessary for someone to recalculate anything should be included in the report. Performance Specification 1 includes a list of items that are to be reported.

Some agencies have additional report requirements or require special formats and tables. Therefore, one should contact the regulatory agency to clarify any such special additions. If the design specifications were certified by the vendor, the certificate of conformance must be included in the report. If the laboratory tests for calibration error and response time were performed by the vendor or a consultant, the test results and calculations must still be performed. To avoid challenges to the data, the original data sheets, signed by the person conducting the tests, should be included in the report submitted to the agency. If the data are photocopied, the copies should be signed by the responsible party attesting that the copies are accurate representations of the originals. For future reference, it is also good to include the strip chart record in the report. This record of the performance of the new system may be helpful for troubleshooting later on.

# References

Brooks, E. F., and Williams, R. L. 1976. *Flow and Gas Sampling Manual.* EPA 600/2-76-203.

Elam, D., and Ferguson, B. 1985. Quality assurance aspects of total reduced sulfur continuous emission monitoring systems. In *Transactions—Continuous Emission Monitoring: Advances and Issues* (J. A. Jahnke, Ed.). Air Pollution Control Association, Pittsburgh, pp. 82–102.

International Standards Organizations (ISO). 1991. *Stationary Source Emissions— Determination of the Mass Concentration of Sulfur Dioxides—Performance Characteristics of Automated Measuring Systems.* ISO Standard 7935. Central Secretariat, Geneva, Switzerland.

Jahnke, J. A. 1984. *Transmissometer Systems—Operation and Maintenance, An Advanced Course.* EPA 450/2-84-004.

Jahnke, J. A., and Aldina, G. J. 1979. *Continuous Air Pollution Source Monitoring Systems.* EPA 625/6-79-005.

Jorgenson, R., and Highberger, E. 1985. Reduced sulfur monitoring: A perspective from Colorado and Wyoming. *Transactions—Continuous Emission Monitoring: Advances and Issues* (J. A. Jahnke, Ed.). Air Pollution Control Association, Pittsburgh, pp. 66–81.

Lindenberg, S. P. 1981. Continuous emission monitor performance specification testing and operation at Coal Creek Station. In *Proceedings—Specialty Conference on Continuous Emission Monitoring: Design, Operation, and Experience.* Air Pollution Control Association, pp. 70–80.

Natrella, M. G. 1963. *Experimental Statistics.* National Bureau of Standards Handbook 91. Washington, DC.

U.S. Environmental Protection Agency (U.S. EPA). 1979. *Federal Register.* U.S. Government Printing Office, Washington, DC. 44 FR 58615 October 10, 1979.

U.S. EPA. 1991a. Standards of performance for new stationary sources—Appendix A—reference methods. *U.S. Code of Federal Regulations.* U.S. Government Printing Office, Washington, DC.

U.S. EPA. 1991b. Standards of performance for new stationary sources—Appendix B—performance specifications. *U.S. Code of Federal Regulations.* U.S. Government Printing Office, Washington, DC.

U.S. EPA. 1991c. Standards of performance for new stationary sources—Appendix B—performance specification 1—specifications and test procedures for opacity continuous emission monitoring systems in stationary sources. *U.S. Code of Federal Regulations.* U.S. Government Printing Office, Washington, DC.

U.S. EPA. 1991d. Standards of performance for new stationary sources—Appendix B—performance specification 1—specifications and test procedures for $SO_2$ and $NO_x$ continuous emission monitoring systems in stationary sources. *U.S. Code of Federal Regulations.* U.S. Government Printing Office, Washington, DC.

U.S. EPA. 1991e. Acid rain program: Permits, allowance system, continuous emissions monitoring, and excess emissions; proposed rule. 56 FR 63001 (December 3, 1991).

**Bibliography**

Brooks, E. F., Flegal, C. A., Harnett, L. N., Kolpin, M. A., Luciani, D. J., and Williams, R. L. 1975. *Continuous Measurement of Gas Composition from Stationary Sources.* EPA 600/2-75-012.

Entropy Environmentalists, Inc. 1983. *Guidelines for the Observation of Performance Specification Tests of Continuous Monitors.* EPA 340/1-83-009.

Federal Republic of Germany, Federal Minister for the Environment. 1988. *Air Pollution Control Manual of Continuous Emission Monitoring.* Bundesministerium für Umwelt, Naturschutz und Reaktorsicherheit, Bonn, Germany. (PO Box 120692, D5300 Bonn 1, Germany.)

Ferguson, B. B., Lester, R. E., and Mitchell, W. J. 1982. *Field Evaluation of Carbon Monoxide and Hydrogen Sulfide Continuous Emission Monitors at an Oil Refinery.* EPA 600/4-82-054.

Franczak, T., and Paprocki, T. 1980. Locating and installing equipment for continuous emission monitoring. *Instrumentation Technology* 11:38.

Gipson, S. G., Tapley, D. W., and Hyatt, J. R. 1981. Performance specification testing as experienced by the Tennessee Valley Authority. Paper presented at the Air Pollution Control Association Meeting, Philadelphia. Paper 81-48.6.

Gregory, M. W., Crawford, A. R., Manny, E. H., and Bartok, W. 1976. Determination of the magnitude of $SO_2$, NO, $CO_2$, and $O_2$ stratification in the ducting of fossil fired power plants. Paper presented at the Air Pollution Control Association Meeting, Portland. Paper 76-35.6.

Jackson, J. A., and Sommerfeld, J. T. 1986. Oxygen stratification in industrial boiler stacks. *J. Air Pollut. Control Assoc.* 36:1238–1243.

James, R. E. 1977. State agency experience with stack monitor performance tests. *Proc. Instrument Soc. Amer.* pp. 57–59.

Nader, J. S., Jaye, F., and Conner, W. D. 1974. *Performance Specifications for Stationary-Source Monitoring Systems for Gases and Visible Emissions.* EPA 650/2-74-013.

Peritsky, M. M., and Wood, R. D. 1981. Extractive flue-gas sampling challenges in-situ methods. *Power.* pp. 48–50.

Polhemus, C. 1981. Performance evaluation testing of compliance monitors. In *Proceedings—Specialty Conference on Continuous Emission Monitoring: Design, Operation and Experience.* Air Pollution Control Association, Pittsburgh, pp. 151–164.

Repp, M. 1977. *Evaluation of Continuous Monitors for Carbon Monoxide in Stationary Sources.* EPA 600/2-77-063.

U.S. Environmental Protection Agency (U.S. EPA). 1978. Traceability protocol for establishing true concentrations of gases used for calibration and audits of continuous source emission monitors. (Protocol Number 1). In *Quality Assurance Handbook for Air Pollution Measurement Systems*, Vol. 3, *Stationary Source Specific Methods.* EPA 600/4-77-027b.

U.S. EPA. 1978b. *Manual 1—Source Selection and Location of Continuous Emission Monitoring Systems—Resource Manual for Implementing the NSPS Continuous Monitoring Regulations.* EPA 340/1-78-005A (PB-283433).

U.S. EPA. 1981. *A Procedure for Establishing Traceability of Gas Mixtures to Certain National Bureau of Standards Reference Materials.* EPA 600/7-81-010.

U.S. EPA. 1982. *Performance Specification 1—Specifications and Test Procedures for Opacity.* Continuous emissions monitoring systems in stationary sources—summary of comments and responses. EPA 450/3-82-025.

U.S. EPA. 1983a. *Guidelines for the Observation of Performance Specification Tests of Continuous Emission Monitors.* EPA 340/1-83-009 (PB-126671/A04).

U.S. EPA. 1983b. *Performance Specification Tests for Pollutant and Diluent Gas Emission Monitors: Reporting Requirements, Report Format and Review Procedures.* EPA 340/1-83-013 (PB-128354/A20).

U.S. EPA. 1983c. *Performance Audit Procedures for $SO_2$, $NO_x$, $CO_2$, and $O_2$ Continuous Emission Monitors.* EPA 340/1-83-015 (PB84-128362).

U.S. EPA. 1985. *Air Compliance Inspection Manual.* EPA 340/1-85-020.

Zakak, F., Siegel, R., McCoy, J., Arab-Ismali, S., Porter, J., Harris, L., Forney, L., and Lisk, R. 1974. *Procedures for Measurement in Stratified Gases.* EPA 650/2-74-086a, b.

# 10

## Quality Assurance Programs For CEM Systems

A quality assurance (QA) program is necessary if a CEM system is to provide quality data on a continuing basis. Without the institution of a quality assurance program, a source's CEM data may not be of adequate quality to meet regulatory requirements and can be easily questioned. A quality assurance program is basically a management program developed to assure that quality control activities are performed. Quality control activities, such as daily calibration and quarterly audits, are the procedures conducted to assure that the data generated from the system are both accurate and precise.

Quality assurance programs are mandated under Appendix F of 40 CFR 60. Similar requirements are included in the 40 CFR 75 acid rain CEM rules, which are somewhat more stringent than those of Appendix F. Although Appendix F applies only to NSPS sources that require CEM systems for compliance purposes, many states have either adapted Appendix F or incorporated it by reference in their regulations or through permit processes. It therefore has a wider applicability than originally stated. This chapter will discuss Appendix F in addition to the basic elements of a CEM system quality assurance program.

### A QUALITY ASSURANCE FRAMEWORK

Ideally, quality control activities should be conducted in all phases of a CEM system project. Beginning with the original purchase specifications, to system performance years after installation, quality control is important. Table 10-1 summarizes the phases of a CEM program in which quality control activities are recommended.

244

**TABLE 10-1    A Framework for Quality Control Activities in the Different Phases in the Source CEM Program.**

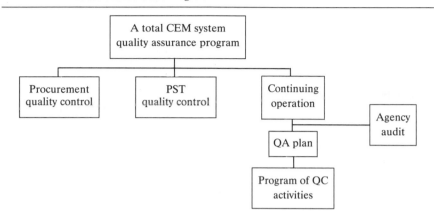

The three important phases in the development of a CEM system are as follows:

1. purchasing the system
2. installing and certifying the performance of the system
3. providing mechanisms for the continuing operation of the system

There are established quality control procedures that can be applied to each of these phases. The source assumes the responsibility for the proper performance of the CEM system and develops a QA program as a means to achieve this performance. An agency may require that such a program be developed and may review and approve the QA plan after it is developed. The role of the plant is then to implement the plan. The role of the agency is to see that the plan is implemented. The agency may inspect or audit the CEM system on a periodic basis to assure, in its own right, that quality data are being generated.

## PROCUREMENT QUALITY CONTROL

A source required to install a CEM system will want to install the "best system" at the "best cost." The "best system" must be considered with respect to limitations of the installation site, flue-gas conditions, and other plant-specific requirements such as work-force limitations and data processing requirements. As has been emphasized in this book, it cannot be stated categorically that any one system is better than another. A system operating well at one plant may be totally unsuited for installation at another. Also, the system offering the "best cost" does not mean "low cost." Purchasing a "low-cost" system may doom a facility to excessive continuing costs for maintenance and repair. Therefore, a balance must be achieved between system quality and cost. For these reasons, CEM system selection and evaluations should always be made on a case-by-case basis.

An orderly procedure should be developed for the procurement of source monitoring systems. An excellent paper by Kopecky and Rodger (1979) details such procedures for ambient air monitors. Some of the ideas presented by Kopecky and Rodger can be applied to CEM system purchases and are discussed here. Basically, there are three areas where quality control should be exercised when purchasing a system:

1. prepurchase evaluation and selection
2. writing the system technical specifications
3. record keeping

Each of these areas will be discussed in turn.

### Prepurchase Evaluation and Selection

Before purchasing a CEM system, an assessment should be made of systems that are currently commercially available. It is possible for the plant to design and construct a system from basic components (analyzers, pumps, probes, etc.); however, this is not often done today. Instead, the plant will consider the systems offered by "turn-key" vendors, who can design, construct, install, and certify the system. The information that has been provided in the previous chapters should assist in the assessment of the literature and sales claims of the various vendors. Such an assessment should include the following activities:

1. comparison of instrument specifications
2. evaluation of operational design with respect to the installation site
3. contact with operators of installed systems
4. field test (if possible—recommended for multiple system purchases)

An ordered program of information gathering and evaluation should be conducted in this step of the process. Vendor literature and relevant publications should be obtained. This information should be reviewed to determine the specifications of the analyzers and systems. These specifications include such parameters as response time, instrument range, temperature limitations, voltage requirements, and so on. If design specifications are mandated for an instrument (e.g., transmissometers), it is important to document that the instrument is advertised as meeting them. If the vendor does not advertise these features, the instrument may not be designed for regulatory applications.

An important activity in reaching a purchase decision is to compare and evaluate the operational design features of different systems. Methods used for monitoring the sample gas (extractive or in-situ), type and length of sample line, conditioning requirements, vulnerability to vibration and temperature, and data processing options should all be considered with respect to the installation site and data requirements.

Current users of commercial systems should also be contacted. Individuals who have been involved in similar monitoring programs should be consulted both for their experiences and suggestions. Although an existing installation may be quite different from the projected one, operating experience from it may be quite useful in an evaluation. Also, numerous papers have been published on CEM system analyzers and their applications, and comparative studies have been undertaken to evaluate different systems.

Quite frequently, a company may wish to standardize, using one CEM system design for all of its plants. Before entering into a major contracts for multiple systems, it is recommended either to conduct a comparative study between systems offered by different vendors or to install one system on an approval basis. In either case, information can be gained regarding a system's suitability for companywide application.

### CEM System Technical Specifications

After soliciting advice, presenting information before management, and meeting to discuss various options, a consensus should be reached as to the desired features of the system (extractive, in-situ, dilution, nondilution, NDIR, NDUV, etc.). At this point, a detailed set of CEM system technical specifications should be prepared. These specifications should then be sent to at least three or four CEM system vendors, who will submit technical and cost proposals for the work.

Writing a complete set of specifications is a crucial quality control activity in the procurement of the system. A good specification provides a basis for the purchaser to obtain the system desired and a basis for legal action if it is not. Either through lacking CEM system knowledge or wishing to expedite a purchase, the plant may merely issue a purchase order or modify a standard instrument specification for this purpose.

Procuring a CEM system in this manner may severely limit recourse if the system proves inadequate for the installation.

A typical CEM system specification incorporates the following:

1. *Purpose.* A brief statement of where the CEM system will be installed, the number of units that will be monitored, and statement of the regulatory requirements applicable to the installation.
2. *Scope of work.* An outline of hardware and services to be provided by the vendor. This section may include a basic system configuration, a list of the number of analyzers required, and data acquisition and control requirements; if desired, it may specify brand names of analyzers or components. Vendor-furnished services may include complete system engineering, installation, and start-up, if desired.
3. *Equipment and services provided by others.* A listing of equipment and services that the vendor is not expected to supply. This may include equipment or supplies such as elevators, ports, catwalks, platforms, electrical supplies, foundations, or calibration gases. Services supplied by plant personnel or others may include system installation, wiring, or certification.
4. *Description of operating conditions.* A description of environmental and stack-gas conditions at the sampling locations. Diagrams of sampling ports and access conditions should be provided here or referred from here to the appendix of the specification. Flue-gas characteristics such as moisture content, velocity, and temperature, and the expected composition and concentrations of pollutants in the flue gas should be supplied. This information is critical to the vendor for the design of the system.
5. *Design criteria and construction.* A detailed description of the system on which the bid is to be prepared. The intent is not to provide all design data, but to provide the vendor with an understanding of the system requirements from both regulatory and operational aspects. Design requirements include adherence to standards, codes, and regulations. They also include specifications for instrument range, drift, and response time. They may include specifications for sample conditioning, interfacing with other plant systems, and data acquisition requirements and reporting formats.

   This section will constitute the bulk of the specification; however, care must be taken not to "overspecify" the system. Vendors must be allowed leeway in the design to use their own experience in CEM systems for the job. If the requirements are too stringent, either no one will bid on the system or they will be ignored in the systems offered.

6. *Vendor-furnished services.* A listing and description of services desired from the vendor. These may include total project management, installation, training, or on-going maintenance services.
7. *Inspection and testing.* A listing of certification guarantees and warranties expected from vendor. These may include factory checkout and certification provisions, performance specification test guarantees, and system availability requirements.
8. *Equipment delivery and requirements.* A statement of progress report requirements, delivery dates, and shipping requirements.
9. *Engineering data and documentation.* A listing of required system documentation. This should include accurate system schematics and wiring diagrams, operating manuals, maintenance instructions, and DAS operating instructions and documentation.

The technical specification may also include specification fill-in sheets, plant schematics, the plant operating permit, and copies of applicable regulations. Care should be taken in preparing specification fill-in sheets, because if they are not designed correctly, the vendor may merely fill in the sheets and not prepare a proper proposal for the job.

The set of technical specifications will normally be accompanied by a set of "standard terms and conditions" prepared by the company contracts department. This document will include legal requirements for insurance, limits of liability, remedies, disputes, and so on. Attorneys prepare these things, but again they should not be made so restrictive that no one would wish to bid on the system.

The Electric Power Research Institute (1984, 1988) has developed a model to use for CEM technical specifications for applications to fossil-fuel-fired utilities. These documents also provide review procedures for the evaluation of vendor proposals.

## Record Keeping

Quality control applied to this phase of a monitoring program includes record keeping. A separate file should be started at the beginning of the evaluation process and should be continued over the lifetime of the system. Vendor literature, phone logs, meeting notes, and financial records should be part of this file. Later, performance specification test data, copies of quarterly excess emission reports, instrument logbooks, and maintenance reports should also be retained. The purpose for this is that, in the case of failure of memory or changeover in personnel, the experience gained in the monitoring project will be retained by the company.

### System Purchase

After the receipt of the CEM system technical specifications, each vendor will prepare a proposal and bid on the project. The evaluation of these bids is central to the purchase of the system. They should be evaluated on their technical merits as well as on cost. If the low bidder is automatically assured of the contract award, the whole technical specification process becomes merely an empty exercise.

## PERFORMANCE SPECIFICATION TEST QUALITY CONTROL

The performance specification test for gas monitoring systems is basically a series of scientific experiments. An unknown physical value (the concentration of pollutant or diluent gas within the flue gas) is determined by two independent methods. The reference method is regarded as a standard, whereas the other, the CEM method, is compared against it. In good scientific practice, quality control activities are performed throughout an experiment. Accordingly, quality control activities should also be conducted during a performance specification test.

Depending upon the complexity of the test, a quality assurance project plan may be developed and incorporated into the test protocol. The preparation of such a plan can help to assure parties involved that the test procedures will be adequately quality controlled. Whether a plan is prepared or not, the following issues and activities should be considered during the test.

### Project Organization and Responsibility

Responsibilities during the test should be clearly defined. The role of the agency observer, plant engineers and environmental specialists, source test team leader, and source testers should be stated. If the organization of the test and activities associated with the participants are clearly stated, less confusion will occur during the test period.

### Sampling Procedures

Procedures to be followed during the testing should be clearly defined. If manual EPA reference methods are being used for the relative accuracy tests, U.S. EPA quality control activities documented in Volume 3 of the *Quality Assurance Handbook* (U.S. EPA 1977a) should be applied.

Other quality control procedures may be followed. For example, if manual reference method tests are required, it may be useful to conduct alternate, automated reference method tests in addition. If a dispute arises between the CEM system data or manual reference test data, the automated method information may help to resolve the issue.

Conditions for representativeness in terms of plant load conditions, sampling conditions, procedures for treating data in case of plant shutdown, and so on should also be addressed. Also, reference method data and CEM system data should be taken at the same time and compared over the same period. Corrections will be necessary for CEM system measurement times for systems having long sampling lines or slow response times.

As mentioned in the previous chapter, time-sharing CEM systems present a significant problem in comparing measurements. If a CEM system monitors for only 5 min over a 15-min period, the data obtained may not be comparable to that obtained by a reference method over a 20-min period. This is a significant issue for sources with rapidly fluctuating gas concentrations. The issue should be resolved between the test participants at the pretest meeting, before the test is initiated.

During the testing, the plant should operate either under normal conditions or at emission levels close to the emission standard. Because of the peculiarities of the U.S. EPA relative accuracy test calculations, the higher the reference method test value (RM), the easier it will be to pass the test. Because of the presence of agency personnel during the test, the plant operator may wish to reduce pollutant emissions to levels lower than normal. Such a practice will make it more difficult to pass the relative accuracy test.

## Sample Custody

Chain of custody procedures should be followed for all manual reference method samples obtained during the relative accuracy tests.

## Calibration Procedures and Frequency

Performance specification test procedures are particularly important for CEM systems. Although Protocol 1 gases are not required for the calibration drift tests, if the automated reference methods [e.g., RM 6C or RM 7E (U.S. EPA 1990a)] are used, Protocol 1 gases are preferred for the reference method instrumentation (EPA 1990a).

A good technique to follow prior to conducting the relative accuracy test is to compare the calibration gases used to span the CEM system with the Protocol 1 (or other) gases used to calibrate automated reference method instrumentation. This comparison will help to determine systematic errors attributable to calibration gas values.

## Analytical Procedures

Laboratory analytical procedures for manual reference method procedures should only be conducted by experienced personnel. It is also suggested that audit samples be used to check laboratory analytical procedures.

## Data Reduction, Validation, and Reporting

All data and calculations should be checked independently by the agency and the organization contracting the source test team (vendor or source). Ideally, raw test data should be entered into computer programs developed specifically for such quality control checks. Because errors frequently occur in these calculations, it is imperative that the calculations be recalculated under any review.

To verify source test data further, it is recommended that, *during the test*, the agency observer and the source representative, together, periodically review strip chart data and computer-generated output. Both should annotate the record, note full-scale values, zero offsets, calibration values, and emission values at that time.

For strip chart traces, a representative emission value should be interpreted and written on the strip chart. Both the agency observer and source representative should initial this value and write down the date and time of the note. For computer-generated outputs, emission values obtained during the relative accuracy test period should be circled and similarly initialed. These procedures can minimize questions that may arise after the raw test data are photocopied or when, long after the test has been performed, recollection of offset and span values has been forgotten.

## Internal Quality Control Checks

The source testing firm should incorporate its own quality control checks in its testing procedures. For automated reference methods, additional instrument verification checks such as linearity and interference checks may be applied.

The source may similarly incorporate internal quality control procedures in the operation of the CEM system. Prior to the test, it may be desired to check a transmissometer with an audit jig and reference filters, or the CEM system may be checked with in-house standards. If the source maintains a portable (or "transportable") gas analysis system, this may be used to check the system performance prior to the relative accuracy test. CEM system voltages, test points, pressures, flow settings, and so on may also be checked prior to the test.

### Quality Assurance Reports

Quality assurance procedures performed should be discussed in the performance specification test report. Any quality control data obtained should be provided either in a separate section or in an appendix to the report. The inclusion of such can assist in checking the overall validity of the test data.

Other quality control procedures may be instituted by the plant or source test contractor when conducting the performance specification tests. The actual performance tests are, however, quite well-defined and, if followed carefully, should produce quality results.

## CONTINUING OPERATION

Successful CEM systems, in practically all cases, are associated with established quality assurance and quality control programs. Concern with system performance cannot end after the system passes the performance specification test, but must be an ongoing commitment for the lifetime of the system. Quality data from a CEM system can be best obtained by instituting a CEM system quality assurance program for continuing operation.

The U.S. EPA has mandated CEM system quality assurance programs for sources that use CEM systems for monitoring compliance with emissions standards (U.S. EPA 1990c). These so-called Appendix F requirements provide direction for the development of a quality assurance plan, calibration procedures, and audit procedures. The primary sections of Appendix F address the following:

1. Quality assurance plan and quality control activities
2. Calibration drift criteria
3. Performance audits
4. QA report requirements

These sections of Appendix F actually constitute a minimum set of requirements for a source QA program. Other procedures can be instituted to assure further the accuracy and precision of CEM system data.

## The QA Plan

A QA plan states the source's philosophy and approach to the quality assurance program. It is important because it establishes the implementation procedures for the quality control activities. These activities actually compose a set of standard operating procedures (SOPs), which are then incorporated into a quality assurance manual that provides CEM system descriptions, company quality insurance policies, and detailed CEM system quality control and audit procedures.

The manual should be a working document and should constitute company policy regarding CEM system performance. Various formats for quality assurance manuals have been developed. One commonly used format is found in U.S. EPA (1976). At a minimum, the CEM manual should include in separate sections, discussions of items given in Table 10.2.

## Quality Control Activities

There are three levels of quality control activity that should be established for a CEM system (for either gas or opacity monitoring systems):

Level 1    operation checks (daily check, observations, and adjustments)
Level 2    routine maintenance (periodic preventive maintenance)
Level 3    audits

*Operation checks* are those procedures that are performed on a routine basis, generally daily, to determine whether the system is functioning properly. These procedures include daily zero and calibration checks and visual checks of system operating indicators such as vacuum and pressure gauges, rotameters, control panel lights, and so on.

*Routine maintenance* is performed at regular intervals. These activities include replacing filters, replacing bearings on motors, reconditioning pumps, replacing lamps, and so on. Depending on the system component, the period may extend from 30 days to a year or more and may have to be determined by trial and error.

*Performance audits* provide a check of system operation that can identify problems, identify the need to improve preventive maintenance procedures, or alert the operator to the need for corrective maintenance.

Activities performed in both quality control levels 1 and 2 are, in large part, preventive maintenance activities. Observations and checks made to see if systems are functioning, as well as calibration adjustments and part replacements, all serve to keep the system from failing.

**TABLE 10-2    Outline for a CEM System QA Manual**

Section 1—The Quality Assurance Plan

1. Quality policy and objectives
2. Document control system
3. CEM regulatory mandates and CEM system description
4. Organization and responsibilities
5. Facilities, equipment, and spare parts inventory
6. Methods and procedures—analysis and data acquisition
7. Calibration and quality control checks
8. Maintenance—preventive
9. Systems audits
10. Performance audits
11. Corrective action program
12. Reports
13. References

Section 2—Standard Operating Procedures

1. Start-up and operation
2. Daily CEM system inspection and preventive maintenance
3. Calibration procedures
4. Preventive maintenance procedures
5. Corrective maintenance procedures
6. Audit procedure 1—cylinder gas audits
7. Audit procedure 2—relative accuracy test audit
8. Systems audit procedures
9. Data backup procedures
10. Training procedures
11. CEM system security
12. Data reporting procedures

Appendices:

A. Facility operating permit
B. CEM specifications and rules
C. Reference test methods
D. Blank forms

*Corrective maintenance* (nonroutine maintenance) is performed when the system or part of the system fails. A failure of the system may just occur, it may be uncovered during an operation check while conducting routine maintenance, or it may be discovered during a performance audit. Experience, along with a well-established preventive maintenance program, should enable the operator to predict failure rates of system components. The number of system failures should decrease if the preventive maintenance schedule is changed routinely to establish component replacement dates consistent with their failure rates. See U.S. EPA (1977c) for a discussion of suggested routine maintenance procedures, typical CEM system problems, and possible corrective actions that can be taken.

### Calibration Drift Criteria

The calibration of CEM systems are checked typically once every 24 h (U.S. EPA 1990c). This may be done either manually or automatically, but is commonly initiated automatically by the CEM system controller or computer. The calibration check is conducted at two levels: at a low-level value (at 0 or at a value between 0 and 20% of the span value) and at a high-level value (typically, between 50 and 100% of the instrument span value). This is normally done for gas and opacity monitors.

The instrument operator should not have to "rezero" or "recalibrate" the system every 24 h when the values are *checked*. Small values for drift may be due merely to system noise. Normally, control chart procedures are developed where drift limits are established for the physical adjustment of the analyzer calibration. In the United States these limits are to be set at twice the PS drift specification. Because the drift specification is 2.5% of the span value for $SO_2$ and $NO_x$ CEM systems, the system does not have to be adjusted until the drift exceeds 5% of the span value. For example, if the span value is 500 ppm, drift adjustments would not be required unless the drift exceeded $\pm 25$ ppm. Of course, the instrument operator may tighten these specifications, if desired.

For CEM systems that automatically correct for calibration drift, unadjusted concentrations must be recorded prior to any automatic adjustments. Also, the amount of adjustment is to be recorded. Many systems merely correct computer-generated data for any amount of drift. This is a computer calculation exercise and involves no physical adjustments of potentiometers or controls on the instrument. The actual adjustment requires action by the operator. In such cases, a computer-generated alarm should be incorporated into the system to alert the operator that action is necessary.

*Out-of-control* conditions for excessive calibration drift may also be defined. For example, in the U.S. EPA Appendix F requirements, if either the low-level (zero) or high-level calibration check result exceeds twice the PS drift specification for five consecutive days or 4 times the drift specification for any one day, the system is said to be out of control. During an out-of-control period, the data are not acceptable as compliance data for the agency. Also, the system is "unavailable" for monitoring and the time the system is out of control cannot be counted in meeting any system availability requirements, where availability (expressed in percent) is defined as

System availability

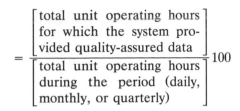

$$= \frac{\left[\begin{array}{l}\text{total unit operating hours} \\ \text{for which the system pro-} \\ \text{vided quality-assured data}\end{array}\right]}{\left[\begin{array}{l}\text{total unit operating hours} \\ \text{during the period (daily,} \\ \text{monthly, or quarterly)}\end{array}\right]} 100$$

Periods of calibration or audit are generally not exempt from the calculation of data availability. However, availability may also be defined as the ratio (expressed in percent) of the CEM system operating time divided by the source operating time or the source operating time minus CEM system calibration time. Availability specifications are quite common in CEM requirements written into source operating permits.

## AUDITS

An audit is a review of the CEM system by a team or an individual not responsible for its day-to-day operation. The principal objective of an audit is to determine how well the system is working. There are two basic sets of audit procedures that can be applied to continuous monitoring systems:

1. the systems audit
2. the performance audit

A *systems audit* is a qualitative evaluation that would normally be conducted by a state agency inspector or a corporate environmental auditor. Here, the operational status of the monitoring system is evaluated and records and data are reviewed. This is not a "hands-on" audit, but an inspection of system operation and system management practices.

A *performance audit* is a quantitative evaluation and is more detailed than a systems audit. It would normally be conducted by a trained agency inspector, a source testing contractor, or plant quality assurance personnel. The performance audit involves testing the system using manual reference methods, alternate automated reference methods, certified cylinder gases, or other audit materials such as calibrated filters or standard solutions. This audit requires a set of audit equipment and materials and is a hands-on activity. Its purpose is to provide a quantitative assessment of problems that might affect the accuracy of the system.

## SYSTEMS AUDIT PROCEDURES

The CEM systems audit can be conducted as part of a corporate environmental audit or as part of an agency source inspection. In both cases, the inspector or auditor is removed from the day-to-day operation of the system and should be able to provide an objective overview. There are a number of ways such an audit can be conducted, but the depth of the audit rests primarily on two factors: (1) the experience of the inspector and (2) the time available to conduct the review.

A CEM systems audit incorporates an examination of all major parts of the monitoring program. This examination includes the following:

1. a tour of the CEM system installation to review the system configuration and condition
2. an evaluation of the CEM system operational status
3. a review of data and records

The audit constitutes a formal, systematic review of the status of the CEM system. The audit is a defined activity and should not be conducted in conjunction with other administrative business at the facility unless it is part of an overall facility inspection. The audit should initiate with an entrance briefing and should continue with a facility tour, data gathering, discussions with plant personnel, and evaluations; it should conclude with an exit briefing. Throughout the audit, the auditor should take notes that are sufficiently detailed to support the findings that are subsequently reported.

If the source has written a CEM system quality assurance manual, the manual can be used as a tool in conducting the review. In fact, such a manual makes the review very easy for the auditor. The auditor can merely go from section to section of the manual, ask if each of the quality control procedures are being followed, and ask for corroborating data documenting that the procedures were indeed performed. Documentation might include inspection sheets, logbook entries, management reports, or test reports. This is where the greatest gaps in the quality assurance program will be found and where suggestions can be made for improving the operation and maintenance of the system.

If the audit is dedicated solely to CEM system inspection and more time is available, an in-depth review of CEM system records may be possible. Cross-checking strip chart data, computer printouts, and excess emission reports can be valuable in developing confidence in reported information.

## CEM System Site Tour

The tour of the CEM system installation is aimed at familiarizing the auditor with the system and its operation. All aspects of the installation, from probe to final data output, should be inspected. The auditor should have a basic understanding of the system prior to the tour, obtained from a review of the agency records, such as the initial performance specification test report, or from excess emission reports. Alternatively, the plant may provide information prior to the audit visit or in the briefing held before the tour.

There are many items to address on the tour, and checklists are often developed to assist new auditors or to refresh the memory of experienced auditors. With respect to the CEM system installation tour, the checklist should provide an assessment that responds to questions such as those listed in the following text. Not all of the items given here will be relevant to a particular CEM installation; however, the example questions and suggestions are provided to assist the auditor in the development of site-specific checklists.

### System Configuration

1. Is the system configured as it was when it received initial approval?
2. Is the system configured as it was when it was certified for monitoring?
3. Are there any modifications to the system that may significantly affect its performance?
4. Are there any major components that have been replaced since the certification or since the last audit? Are the analyzers the same (check serial numbers)?

### System Condition—Stack or Duct Installation Points
(*Note*: Request plant personnel to open protective coverings, instrument cabinets, etc., when feasible.)

1. What are the access and site conditions?
   a. Is the access to the site difficult? Is the elevator working? Would maintenance personnel be willing to inspect the site once a day? Once a week? Never?
   b. Is the site protected? Between liner and chimney? On the catwalk? Can maintenance personnel perform repairs at the site during inclement conditions?
   c. Does the site show evidence of being visited? Is there dust or fly ash on the railings or hand holds?
   d. Is the temperature higher or lower since the last audit?
   e. Is vibration at the location higher or lower since the last audit?
   f. What is the stack static pressure?
   g. Are there puddles of water at the site? Does it rain on the apparatus, or does water drip onto the system?
   h. Has fly ash sifted into piles near the installation.
   i. What is the condition of the ambient air? Is it laden with blowing fly ash? Is the odor of $SO_2$ or other gases present? Are ambient pollutants affecting the CEM installation?
2. How does the probe or monitor installation look?—Clean? Dirty? Well maintained?
   a. Do significant quantities of dust or fly ash cover the unit?
   b. Are bolts rusted tight on the mounting flange? Is there evidence that the probe or unit has been recently removed?
   c. Is the unit corroded?
   d. What is the condition of plastic or rubber components such as gaskets, heat-traced line protective coverings, hoses, electrical cables, and so on?
3. How does the probe itself look? (sometimes it may be possible to look at the probe through another port.)
   a. Is it black, obviously impacted with particulate matter?
   b. Is agglomerated, sticky particulate matter adhering to the probe?
   c. Is the probe sagging?
   d. Is the probe oscillating with the stack flow?
4. What is the condition of stack-mounted instrumentation?
   a. For blowers (in-situ gas monitors or transmissometers), are the filters clean? Is a supply of air being provided to the windows?
   b. Are any protective shutters tripped?
   c. What is the status of any warning lights or operation indicators at the site?
   d. For junction boxes or meters at the site, record meter values and note times to correlate the information with previous records or data recorded in the control room.

5. Are calibration gas or audit gas cylinders located at the probe site?
   a. If so, record cylinder number and tag information (ppm/% values, type of certification, gas vendor, date of analysis). What confidence is there in the stability of cylinder gases used for more than a six-month period? More than a 1-yr period?
   b. Record regulator pressures for each cylinder—tank pressure and supply pressure. Also note original tank pressure.
   c. Are the regulators corroded? Are proper fittings used for acid gases?
   d. What is the condition of the gas supply lines and fittings? Is the same type of tubing being used as at the time of the last audit or the PST? Are the fittings mangled or corroded, or do they show other indications of abuse?
6. If possible, observe a gas calibration cycle and a probe blowback cycle at the installation site.
7. For transmissometer systems, note the following:
   a. alignment (check alignment sight)
   b. condition of purge air hoses (cracked, loose)
   c. condition of purge air motor (motor noise, vibration)
   d. condition of purge air filter system
   e. condition of electrical cables
   f. status of shuttering system
8. What else can be observed? Is garbage or paper strewn about the site? Are there tools, instrument manuals, and source test apparatus at the installation?

### System Condition—Umbilical Lines and Electrical Cable

Note the following when walking from the probe stack or duct installation to the CEM system analyzer shelter or CEM system room.

1. Is there a minimum slope of 5° from the probe to the conditioning system?
2. Are there any loops or kinks in the umbilical line?
3. Does the umbilical coil back on itself or touch itself or another umbilical at any point?
4. Are there any unheated sections (e.g., where two umbilicals are spliced together, or just after the probe assembly, or just before the conditioning system)?
5. Are electrical cables properly routed and protected?
6. Are the cables located near or bundled with power lines, near electric motors or equipment generating strong electromagnetic fields?

7. For both heated and unheated line, what is the condition of the line from the port to the conditioning system? Is it corroded, brittle, or dirty? Has it been damaged, spliced, or otherwise repaired?

## Auditing in the CEM Shelter or Control Room

The CEM system installation tour will most likely end (or begin) in the CEM room, CEM shelter, or plant control room. This room generally contains the sample conditioning system (for some extractive systems), analyzer control panels, and/or analyzers.

The auditor should determine the proper methods of the system's operation from either the manufacturer's written instructions or, preferably, the standard operating procedures provided in the plant quality assurance manual. At this point, it is often convenient for the plant environmental engineer or CEM system operator to explain entries in the instrument maintenance log and describe the routine system maintenance procedures. Maintenance activities related by plant personnel should be cross-checked in the maintenance log and quality assurance manual to determine if established procedures are being followed.

For the part of the audit conducted in the CEM system shelter or control room, the audit checklist should provide an assessment that responds to items such as those listed next.

### System Condition—Conditioning System (For Extractive Systems)
1. Trace the path of the umbilical line or sample line to the conditioning system or analyzers. Attempt to understand how the system works at this point. In particular, note the following:
   a. Does sample gas enter the conditioning system under pressure or under vacuum?
   b. What is the technique used for removing water from the sample gas?
   c. How is condensed water removed from the conditioning system? What is the probability of pollutant gas being absorbed in the condensed water? Is algae growing in the coolant?
   d. Locate the drain pipe for the condensed water. If the pipe drains to the outside, can the outlet freeze up during the winter?
   e. Does the system incorporate a moisture breakthrough sensor? Does it work?
   f. Can condensed liquid be seen in any of the Teflon lines?
   g. Is the tubing neatly arranged or haphazardly? Are there indications of ad hoc plumbing changes?
   h. Are the fittings, valves, and so forth corroded or leaking?

i. What is the status of the fine particulate filter? Is it clean or dirty? How often is it replaced? When was it last replaced?
j. What is the condition of the sample pumps? Are they corroded, noisy, leaking? When were the diaphragm and bearings last replaced? (Verify in the maintenance log.)
k. Note system gas flow (rotameter) readings and pressure readings. Are they consistent with readings obtained during the last audit? Are they consistent with readings specified in the operating or quality assurance manual?
l. Is a manifold used for gas distribution to different analyzers?
m. In flow-through systems, where does the dump line exhaust?

### System Condition—Monitor Control Panels and Monitors

1. What is the status of the control panel lights and indicators, alarms, and so on for each analyzer?
2. Record panel meter readings exhibited by each analyzer.
3. Record settings for zero and span control dials for each analyzer (for comparison with historical data).
4. Record values for panel-operated test parameters such as lamp voltages, automatic gain controls, and so forth (if procedures do not interfere with analyzer operation and data recording).
5. For transmissometer systems, ask the operator what the stack exit correlation factor is and how it is set in the transmissometer. Also ask if the nominal value of the filter has been corrected by this factor (see Chapter 7). Follow with other questions such as these:
   a. Was the flange-to-flange distance used to determine the transmissometer path length? (The inner stack diameter should be used.)
   b. How was the stack exit diameter determined?
   c. Was a single-pass value or a double-pass value used for the transmissometer path length? (Which one is used depends on the type of transmissometer.)
   d. Verify the appropriateness of the internal span filter value. Look up the specified value in the appropriate regulation, and determine if a filter with the correct value has been installed.

### System Operational Procedures

1. Who has responsibility for
   a. system operation?
   b. calibration?
   c. system preventive maintenance?

    d. system corrective maintenance?
    e. auditing?
    f. reporting?
2. Ask questions such as the following:
    a. How many hours a week are spent operating and maintaining the system?
    b. What has given the most trouble since the last audit?
    c. How good has the CEM system vendor been in emergency service response? In spare parts delivery?
3. Ask the operator to perform a system calibration. Note the following:
    a. familiarity of the operator with procedures
    b. how the operator determines that the system is in calibration (which data are used—meter, strip chart, or computer output?)
    c. how the operator annotates the data (in a logbook? on the strip chart? on the computer output? not at all?)
    d. how it is determined that the system is or is not in calibration (are quality control charts used?)
4. Record the calibration data and obtain copies of the strip chart record and/or computer output, if possible.
5. Either at this point or when more time is available, cross-check observed procedures with written procedures given in the plant CEM system quality assurance manual.

### Review of Records and Data

Records and previous data should be reviewed in a location that is free from interruptions and is relatively comfortable. Several hours should be devoted to this activity. Depending on the complexity and condition of the CEM installation and on the skill and patience of the auditor, the review may take even longer.

If the source has developed a CEM system quality assurance plan, the plan can be used as a tool in conducting the review. The auditor should proceed form point to point of the plan, determine if each of the quality control procedures is being followed, and ask for corroborating data documenting that the procedures were performed in the last quarter or the last year. Documentation might include inspection sheets, logbook entries, management reports, or test reports. This is where the auditor will find the greatest gaps in the quality assurance program and where suggestions can be made for improving the operation and maintenance of the monitoring system.

Review the logbooks carefully for frequency of corrective maintenance, parts replacements, and any unusual occurrences. For gas monitoring systems, note change-out dates for gas cylinders and note entries for new gas concentrations. For transmissometer systems, note the frequency of filter changes and the frequency of window cleaning and lamp replacement. Also note any recurring failures. If there are many entries, there should be some concern over the capabilities of the system. It has been documented (McCoy 1990) that transmissometer systems can operate with an availability of better than 97% of the time and gas monitor systems exhibit 90–95% availability. Downtime greater than 5% should merit further investigation.

Daily, weekly, or monthly inspection forms should be reviewed for completeness and consistency with system logbooks. Review the inspection reports for historical consistency. If meter or gauge values are entered in forms, note any changes occurring over the review period.

Other records, such as reports of quarterly internal quality assurance audits, performance audits, off-stack zero checks, or repeats of calibration error or drift tests, should also be examined.

The U.S. EPA requires that a source retain data records for a period of two years. It is recommended that the data record for the past 30 days be reviewed and that at least three months of maintenance records be reviewed. If possible, the data record should cover a period for which the last summary report has been submitted to the agency. Review data over the period in which any excess emissions were reported. Note if the data correspond and if any annotations are provided, and determine the zero and span values obtained closest to the time of any exceedances. When reviewing the data record, also be observant for the following:

1. missing data
2. unusually noisy or "flat" data
3. inconsistent trends in readings
4. annotations for monitor and source downtime
5. annotations for exceedances
6. printed fault or warning codes

If a CEM system has a strip chart recorder, it is usually used to record opacity values from the transmissometer. A glance at the strip chart record can reveal all types of things and gives a good indication of how well the system is being maintained. If a computer printout is all that is available, the audit will be more difficult. It is harder to detect trends, noise, and other system problems from digital data, although system fault codes are often helpful.

The strip chart record also gives many clues to the quality of the source monitoring program. If the pen is not inking, if the paper is jammed, or if the chart speed is too slow to provide intelligible data, concern should be expressed over the adequacy of the data. A working strip chart should be properly annotated and the record or the daily zero and calibration should be easily read.

Evaluating a computer record will require assistance from the CEM operator. It is difficult to examine a printout (or even to obtain one) without some assistance. When reviewing the data, look for the zero and span values first. This will provide some orientation to the format. If there are any printed fault codes, inquire into their meaning. It may be useful to have the codes pointed out in the data-acquisition system documentation.

The instantaneous strip chart data and the data on computer printouts should correspond. After a calibration check, many computer systems automatically correct subsequent data. This is a computer correction and does not involve the adjustment of the analyzer zero or span potentiometer. In such cases, the strip chart data and analyzer meter readings will not be corrected and will differ from the printed computer-adjusted readings.

### The Audit Report

The audit report organizes and coordinates in a usable manner the information gathered during the audit. It is the compilation of factual information and professional judgment resulting from the audit. Information in the report must be accurate, relevant, complete, objective, and clear. The report also serves to record the procedures used in gathering the data and gives factual observations and evaluations from the audit.

It is imperative that the report be clear and concise. A discussion of general topics should be avoided and all compliance issues should be directly linked to regulatory requirements.

The purposes of the systems audit are to determine how well the system is operating and to obtain some assessment of the quality of the reported data, without conducting any independent performance tests. A well-conducted systems audit will give plant management and/or the agency an indication of how the CEM program is operating and an indication of the level of confidence that can be given to the data. For more quantitative assessment of data quality, the *performance audit* can follow from the systems audit or as part of a regular schedule of audit activities.

## PERFORMANCE AUDIT PROCEDURES

The main idea behind the performance audit is that it provides an independent assessment of the monitoring system accuracy. Daily calibration drift determinations and routine maintenance do not necessarily guarantee that data will be accurate. However, an independent assessment using an appropriate auditing technique can provide an indication of data validity.

Performance audits are conducted either on a regularly scheduled basis as part of a quality assurance program or when the findings of the systems audit indicate that information of a more quantitative nature is necessary to evaluate data accuracy. The most common method of conducting a performance audit of a gas monitoring system is to challenge the system with certified audit gases. The cylinder gas audit (CGA) does provide an independent assessment of a system, but does not always assess the accuracy of the data. Using optical filters for electro-optical analyzers or transmissometers, using standard solutions for electrochemical systems (typically ion-selective electrode analyzers), or challenging the system with known electronic signals can provide other assessment levels. These latter methods are commonly used in Europe, but have been discouraged in the United States, which maintains a preference for cylinder gas audits.

Independent techniques of performance auditing involve measuring the emissions by using a different method. These methods include the following:

1. repeating the certification test
2. conducting abbreviated relative accuracy tests using manual or automated reference methods
3. testing with instrumented mobile vans
4. testing using portable inspection monitors

Such testing is of course more involved and expensive than a systems audit, requiring additional resources.

The choice of performance audit procedures depends upon regulatory requirements and the resources of the owner. The U.S. EPA Appendix F requirements specify that a performance audit be conducted at least once each quarter, using one of the following:

- Relative accuracy test audit (RATA)—a repeat of the relative accuracy test procedures as defined in 40 CRF 60 Appendix B Performance Specifications
- Cylinder gas audit (CGA)—a challenging of the monitoring system with cylinder gas of known concentration (NIST-traceable gases)
- Relative accuracy audit (RAA)—an audit similar to the RATA, except that only three sets of measurement data (instead of nine) are taken

Appendix F requires that at least one of the quarterly audits be a RATA, and either the CGA or RAA can be used for the other three quarters.

Other regulations may increase the frequency of audits or may specify that only certain types of audits may be conducted.

The quarterly audits are intended to alert the CEM system operator to problems that might be occurring with the system. Many problems can develop that may not be indicated by the day-to-day calibration check routines. If excessive inaccuracies occur for two consecutive quarters, as evidenced by the audits, Appendix F requires that the CEM quality control procedures be revised.

### The Cylinder Gas Audit

The cyclinder gas audit is conducted by challenging the CEM system with an audit gas that is traceable to a certified gas. In the United States it is required that audit gases be traceable to a National Institute of Standards and Technology (NIST) standard reference material (U.S. EPA 1977c). The audit gases should be introduced at the sampling probe and should pass through all components of the sampling system.

The method used to introduce the gas at the probe depends on the design of the probe and the monitoring system. Techniques for probe calibration checks ("probe cal") for extractive systems are illustrated in Figure 10-1a and b. For source-level extractive systems, gas may be injected at the probe for either an external filter (Figure 10-1a) or an internal filter (Figure 10-1b) design. The cylinder-gas flow rate is increased until a maximum, stable reading is obtained. Audit gas must be provided at a rate necessary to overcome the absolute static pressure of the stack gas at the probe. At this pressure, stack gas will be flushed away from the probe so that an undiluted audit gas concentration can be obtained. Excess gas is exhausted into the stack as the extractive system withdraws sample through the probe at its normal flow rate. A probe calibration can be conducted at an external filter if a probe sheath is used. However, large volumes of gas may be necessary to flush out the stack gas. A somewhat more effective method is to flood the annulus around an internal probe filter, as shown in Figure 10-1b. This may require less calibration gas because the space is more confined.

It is relatively easy to conduct a probe CGA for a dilution probe because the inner space of the probe can be flooded with audit gas (see Figure 3-15). Because this space is smaller than that encountered in most fully extractive systems, the volumes of audit gas required are correspondingly smaller.

The techniques just discussed have been referred to as *probe vent audit* techniques (Reynolds 1989). Another technique, termed the *external atmospheric vent audit* technique, is used when the probe system cannot be

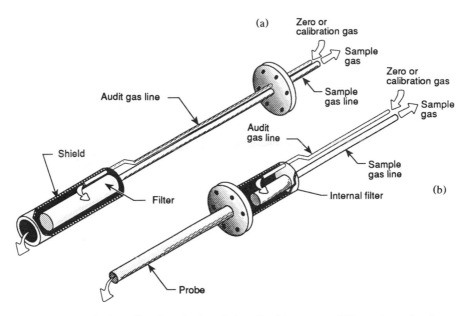

**FIGURE 10-1.**    Probe calibration check technique for (a) an external filter using a sheath and (b) an internal filter.

flooded with audit gas. This less complete check of the probe system can be conducted by using a three-way valve and a rotameter as shown in Figure 10-2. In this method, the probe system is essentially shut off and audit gas is vented through a rotameter. The sample system withdraws the audit gas, at atmospheric pressure, at the valve. Care must be exercised to supply enough gas to vent through the rotameter, otherwise ambient air will dilute the sample. If a three-way valve is not part of the installed system, the auditor may ask the CEM system operator to disconnect the sample line at the probe and connect the audit sample line and rotameter directly to the line. This technique obviously does not check the probe and probe filter, but can assist in evaluating the integrity of the sample line. For CEM systems that pull gas through the sample line under vacuum (negative pressure), the check should not be conducted without the vented rotameter. The system will otherwise be pressurized and the audit will do little to check for leaks in the line.

Performing a cylinder gas audit on a point in-situ analyzer is similar to conducting a probe vent audit for an extractive system. Audit gas can be used to flood the sample chamber to a pressure greater than the stack static pressure, as shown in Figures 6-12 and 6-13. It is possible to overpressurize the chamber in analyzers that incorporate a ceramic filter

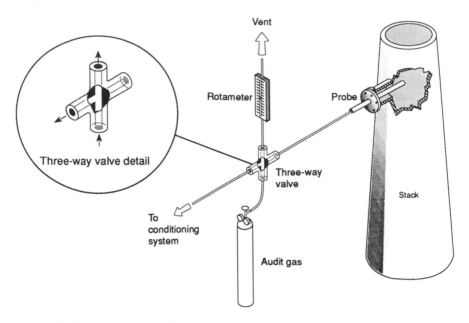

**FIGURE 10-2.**    Audit method for probes that cannot be flooded: the "external atmospheric vent" technique.

to prevent fouling from particulate matter. At higher pressures, the audit gas concentration will be higher than true and will lead to incorrect readings. The audit gas flow rate should be increased 0.5 l/min over the manufacturer's recommended value, to determine if such an effect is occurring [see Reynolds (1989) for detailed procedures].

A cylinder gas audit can also be conducted on an in-situ path analyzer if a flow-through gas cell is incorporated into the measurement system. Figure 6-9 illustrates one such design. In a double-pass system, the zero mirror will reflect the measuring light beam through the gas cell, back to the detector. The flow-through gas cell serves as a sort of "pseudostack." Cylinder gases at higher concentration are necessary (on the order of percent instead of parts per million) in order to obtain an optical depth that corresponds to that of the measurement path length. For example:

- If the analyzer monitors a gas concentration of 600 ppm $SO_2$ over a total distance of 10 m in a stack, the optical depth will be $600 \times 10 = 6000$ ppm-m.

- If a flow-through gas cell has an internal measurement path of 1 cm (0.01 m), a cylinder gas would have to have a concentration of $6000/0.01 = 600,000$ ppm for equivalent optical depth. The 600,000 ppm, of course, corresponds to a concentration of 60%.

For in-situ analyzers, special attachment audit devices (as previously shown in Figures 6-8 and 6-10) can be used to gain additional audit information. For example, in Figure 6-8, multiple flow-through cells can be stacked together. One lower-value audit gas can then be used to obtain multiple audit readings as each cell is filled in sequence.

Meaningful cylinder gas audits on single-pass in-situ path monitors are more difficult to perform. Because the stack gas is in the light path and is being measured, a flow-through gas cell will give an incremental reading added onto the stack-gas values. Data resulting from such measurements become difficult to interpret when the stack-gas concentrations are rapidly varying. A method of avoiding this is to place a pipe across the stack, between the transmitter and the detector. If the pipe can be closed off and flushed with ambient air, a zero baseline value can be obtained. Then the instrument response will correspond more accurately to the audit gas value. However, other variables such as contaminated zero air or system misalignments caused by sealing the pipe may affect the instrument's response.

Another technique that has been used in single-pass units is to block off the measurement light beam and use an auxiliary light source to send a beam through the internal audit cell. Obviously, this scheme does not check all of the normally operating active components of the system. However, it will serve to check the detector and primary electronics.

CEM system vendors frequently state that in-situ path analyzers do not meet Appendix F requirements. This is not true. Path analyzers that incorporate flow-through cells can be challenged with audit gases and meaningful information can be obtained from the results.

A cylinder gas audit is typically conducted (U.S. EPA 1990c) by challenging the CEM system with two audit gases: one having a value 20–30% of span, the other with a value of 50–60% of span. The CEM system is challenged alternately, three times with each audit gas. Sufficient time should be allowed for each injection until the concentration reading stabilizes. For integrating instruments, it may be necessary to monitor several measurement cycles to determine when the reading is stabilized. Audit readings must not be included in the facility emission averages. The CEM system operator should place the data-acquisition system in an alternate mode while the audit is being conducted.

The accuracy calculation for the CGA is given by

$$A = \frac{d_m - c_a}{c_a} \times 100 \qquad (10\text{-}1)$$

where $A$ = CGA accuracy of the CEM analyzer (in terms of parts per million or in percentages)

$c_m$ = average analyzer response during the audit

$c_a$ = certified value of the audit gas

The CEM system is "out of control" if $A$ exceeds $\pm 15\%$. The out-of-control period begins at the time of completion of the audit.

An auditor may wish to modify these procedures in order to obtain further data on the system. Also, state or regional quality assurance guidelines may require other procedures (Peeler 1990). The following are typical variations:

1. using a zero gas in addition to two audit gases
2. using an audit gas at a concentration corresponding to the emissions standard
3. using an audit gas corresponding to the average stack gas concentration of the pollutant
4. generating variable gas concentrations using a dilution system
5. challenging the system only twice with each gas
6. challenging the analyzer with audit gas at the analyzer calibration port

Variations 1 and 6 are particularly useful in trouble shooting a CEM system. It should be noted that a cylinder of zero gas may not be necessary because, for example, an $NO_x$ audit gas reading may serve as a zero reading for an $SO_2$ channel. Using audit gas concentrations other than those specified in Appendix F (40 CFR 60) is either a matter of different regulatory requirements or technical preference [e.g., Butler (1987); Butler and Willenberg (1990)]. Using an audit gas corresponding to the average pollutant concentration readings can give additional confidence in the emission measurements, if the audit results are satisfactory.

Challenging an analyzer at its calibration port (instead of at the probe) is a simple method of checking the audit gas against the span gas used to calibrate the analyzer. If the audit gas reading obtained on the analyzer differs widely from its certified value, it may indicate that the span gas has deteriorated or its tag value is incorrect. If the probe audit results are unsatisfactory, it may be useful to check the audit gases at the analyzer to aid in troubleshooting the system.

*Audit Gases*

Cylinder gases used for calibrating and auditing continuous emission monitors have improved significantly over the past 20 years. In the early 1970s, it was not uncommon to find pollutant gas concentrations differing from the tag values by over 10%, even when the accuracy was stated by the producer as being between 2 and 5% of the pollutant value. This, of course, led to considerable uncertainty in CEM system measurements and often led to the failure of initial performance specification tests. Because accurate standards are needed in any measurement program, steps were taken by the U.S. EPA to improve the quality of cylinder gases.

TABLE 10-3    Differences between Standard and Certified Gases

| SRM | Standard reference material | Prepared and sold by NIST |
|---|---|---|
| CRM | Certified reference material | Prepared by the gas vendor<br>Referenced directly to an SRM<br>Nominal concentration within $\pm 1\%$ of an SRM concentration<br>Analyzed after preparation<br>Analyzed again, 30 days after first analysis<br>Two samples analyzed by independent laboratory |
| GMIS | Gas manufacturer's intermediate standard | Prepared by the gas vendor<br>Referenced to an SRM or CRM<br>Nominal concentration within 0.3 and 1.3 times the concentration of the SRM<br>Must be assayed three times over a three-month period<br>Assays must agree to within 1.0%<br>Must be recertified every three months |
| Protocol 1 gas | | Prepared by the gas vendor<br>Referenced to an SRM, CRM, or GMIS<br>Nominal concentration between 0.3 and 1.3 times the concentration of the SRM, CRM, or GMIS<br>Assays must agree to within 1.5%<br>Reactive gases must be re-assayed after seven days |

In 1978, the U.S. EPA published a traceability protocol that could be used to reference a prepared cylinder gas to National Bureau of Standards [(NBS) now NIST] standard reference materials (SRMs). The protocol was subsequently revised in 1987 [see U.S. EPA (1977d)]. Gases prepared under this protocol are known as Protocol 1 gases. Due to the limited supply and high cost of SRMs, intermediate standards may also be used by the gas vendor when preparing Protocol 1 gases. An intermediate standard may be either a certified reference materials (CRMs) [see Hughes (1981)], or a gas manufacturer's intermediate standard (GMIS) [see U.S. EPA (1977d)]. Table 10-3 summarizes the differences between these standards.

All of the gases are assayed using analyzers that have been calibrated with certified gases. Calibration procedures, linearity response, and adjustment procedures are strictly defined in Protocol 1. When purchasing a Protocol 1 gas, the buyer should obtain a certificate from the vendor that contains the mandated certification data (U.S. EPA 1977d).

The certification applies for a maximum period of 18 months for most gases contained in aluminum or stainless steel cylinders. For other cylinder materials and for nitrogen dioxide–air mixtures, the certification period is only six months. Also, if the cylinder gas pressure should decrease below 100 psi (700 kPa), the Protocol 1 gas should not be used. If sufficient gas remains after 18 months, the gas may be recertified, but the recertified value must be within 5% of the original certified value.

Europe has no standards organization comparable to NIST that can provide certified reference gases for source monitoring applications. However, the International Standards Organization (ISO) has prepared detailed standards for the preparation of calibration gas mixtures (ISO 1981, 1984, 1986).

## Transmissometer Audits

Audit devices (or "audit jigs," as they are frequently called) have been developed for checking the performance of double-pass transmissometer systems (Cohen and Ross 1980; Hamil 1980). Basically, the devices consist of a slot for holding calibration filters and short-range retroreflector assembled into a holder that can be attached onto the transceiver. The device and transceiver basically constitute a "minitransmissometer" that can accommodate audit calibration filters. Figure 10-3 shows a device developed by Hamil (Hamil 1980) for the Erwin Sick transmissometer.

A reflector mirror is contained at the end of the device, serving as a reflector intermediate in position between the simulated zero reflector and

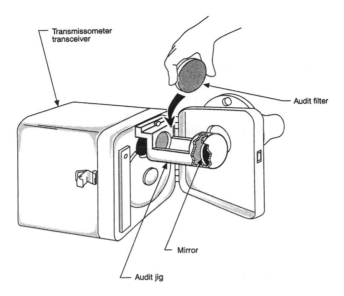

**FIGURE 10-3.**    A transmissometer audit jig attached to the transceiver assembly.

stack retroreflector. Certified filters can be placed between this retro-reflector and the transceiver head to check the calibration of the instrument over a range of opacities. The device also contains an iris, which allows the reflectance of the mirror to be adjusted so that it will correspond to the true, stack zero. This "audit zero," of course, will not necessarily be identical to a clean-stack zero, but contrasting it to the instrument internal "simulated zero" provides a good quality control check.

In the jig shown in Figure 10-3, a cylindrical holder fits over the cast metal lip that holds the transceiver window. A retroreflector is positioned in back of the device; again, the design allows an auditor to slip in different filters to check the instrument calibration.

## Conducting the Transmissometer Audit

Detailed guidance for conducting a transmissometer performance audit, can be found in Plaisance and Peeler (1987). Preliminary to installing the jig, stack exit correlation values, fault indicators, and internal zero and span values are checked, and the transmissometer windows must be cleaned. The actual performance part of the audit is relatively simple. First the audit jig is attached to the transceiver, and a simulated zero value is obtained by adjusting the iris on the jig. This zero value should be the same as the pseudozero established with the transceiver zero mirror. Each of the three filters is then placed into the jig for a period of 2 min each. Stable values are recorded for each filter from the instantaneous (unaveraged) response of the transmissometer. This procedure is repeated five times. The calculations for the calibration error are identical to those used to determine the calibration error in PS 1. The transmissometer system is viewed as passing the performance audit if the calibration error (Er) for each filter is less than 3% opacity.

The transmissometer performance audit calibration error check does not check the absolute accuracy of the transmissometer system. There are many other factors involved in the opacity measurement, such as system alignment and the viability of the cross-stack zero. The procedures of the performance audit, however, point out problems that might affect opacity measurements. Many of these problems are correctable and the audit report should include recommendations for such corrections.

The calibration error check, using the audit jig, is one of the best quality control procedures that can be incorporated in a source transmissometer system quality control plan. Performed quarterly, it can be carried out relatively rapidly and can serve to increase confidence in the opacity data.

### Repeating the Certification Test

As mentioned earlier for gas monitoring systems, the cylinder gas audit is not completely independent of the monitoring system. In fact, most CEM installations can easily pass EPA-established criteria for the CGA. In most cases the exercise is merely a comparison between the audit gases and the system span gases, particularly if the daily calibration check is conducted at the probe.

Repeating the certification test is the ultimate audit technique, because the regulatory acceptance of the CEM system is based on the results of this test. In the United States this is required once a year for sources regulated under Appendix F. Under certain conditions, the certification test is required to be performed twice a year for sources participating in the acid rain allowance trading program.

For sources regulated under Appendix F, this recertification is called the relative accuracy test audit (RATA) and is identical in procedure to the PS 2 relative accuracy test. The CEM system is said to be *out of control* when the relative accuracy is greater than 20% of the average reference method value or greater than 10–20% of the emissions standard (depending on the value of the emissions standard). If the CEM system is out of control, corrective action must be taken and the RATA must be redone to determine if the system is again operating properly.

### Testing with Abbreviated Relative Accuracy Tests

An audit can be conducted by reducing the number of runs that are normally conducted in a relative accuracy test. Instead of performing nine runs, as is required in a reference method, only three runs are necessary in a relative accuracy audit (RAA). The effect of reducing the number of runs on the statistical acceptability of the data has been discussed in detail by Jernigan (1986). Reducing the number of runs can save time, but not necessarily a great deal of expense. Because much of the cost associated with a stack test or audit program is associated with travel, quality assurance, and reporting, a few hours saved on the site will not save much money.

The RAA calculation differs from that of the RATA by the elimination of the confidence coefficient term. The contribution of the confidence coefficient to the RAA is essentially compensated by reducing the out-of-control criteria from 20% in the RATA to 15% in the RAA.

During the out-of-control periods determined by an Appendix F audit, data may not be used for emission compliance purposes nor does the time out of control count toward system availability. These restrictions are the same as those given for excessive calibration drift. Table 10-4 summarizes the out-of-control conditions discussed in this chapter.

### Testing with Portable Inspection Monitors

A number of portable inspection monitors are commercially available. The monitors are generally lightweight, are very portable, and operate using electrochemical cells. Oxygen analyzers in this class are particularly useful and can provide accurate short-term measurements. Pollutant gas analyzers often have relatively long response times (5–15 min) and may drift on a long-term basis. Such analyzers can be useful in uncovering stratification problems or in helping to resolve gross differences between installed CEM systems and manual test methods.

**TABLE 10-4    Summary of Appendix F "Out-Of-Control" Conditions**

| | |
|---|---|
| CD > 2 × PS drift specification | For five consecutive days |
| > 4 × PS drift specification | For any one check |
| RATA > 20% | |
| or | |
| > 10% of the standard | |
| (whichever is greater) | |
| > 15% of the standard | For standards between 130 and 86 ng/J |
| | (0.30 and 0.20 lb/$10^6$ Btu) |
| > 20% of the standard | For standards below 86 ng/J |
| | (0.20 lb/$10^6$ Btu) |
| CGA > ±15% | |
| RAA > ±15% | |
| or | |
| ±7.5% of the emission standard | |

## Testing with Instrumented Mobile Vans

Most commercial stack testing companies and a few corporate environmental organizations today use instrumented test vans or trailers to perform CEM relative accuracy tests and audits (Peeler and Deaton 1981; Chapman 1990; Noland and Reynolds 1990). These vans usually incorporate instrumentation equivalent in quality to that of installed CEM systems. Data obtained from such systems can be of high quality if the requirements of the automated reference methods such as 6C and 7E are met.

## The Performance Audit Report

The results of an Appendix F quarterly audit are reported in a data assessment report (DAR). The DAR is submitted to the agency quarterly. As a minimum, the DAR must contain the following information:

1. source owner name and address
2. identification and location of monitors
3. manufacturer and model number of each monitor
4. audit accuracy results; accuracy results for audits determining that the system is brought back in to control
5. results from reference method performance audit samples (if manual reference method tests were performed)

6. summary of corrective actions taken when system was determined to be out of control

A simple DAR format is given in Appendix F (40 CFR 60).

Appendix F (40 CFR 60) has proven to be useful both to agencies and to sources required to implement quality assurance procedures and quality control techniques. Walsh and von Lehmden (Walsh 1989; von Lehmden and Walsh 1990) have summarized the results of Appendix F audits for Subpart Da electric utility steam generating units for 1988 and 1989. Ninety-six percent of the $SO_2$ RATA data were found to be within the 20% control limit during this period. These data are encouraging and attest to the effectiveness of quality assurance programs in maintaining CEM system reliability.

**References**

Butler, A. T. 1987. A method for quick validation of continuous emission monitoring systems. In *Proceedings—Transactions of the Pulp and Paper Institute (TAPPI) 1987 Environmental Conference, Atlanta*. pp. 199–208 (Paper 12-1).

Butler, A. T., and Willenberg, J. M. 1990. A method for quick validation of continuous emission monitoring systems. In *Proceedings—Speciality Conference on Continuous Emission Monitoring: Present and Future Applications*. Air and Waste Management Association, Pittsburgh, pp. 350–359.

Chapman, J. 1990. Examination of a transportable continuous emission monitoring system. In *Proceedings—Specialty Conference on Continuous Emission Monitoring: Present and Future Applications*. Air and Waste Management Association, Pittsburgh, pp. 327–337.

Cohen, J. B., and Ross, R. C. 1989. Use of precalibrated optical density filters in the in-situ calibration of opacity monitors. Paper presented at the Air Pollution Control Association Meeting, Montreal. Paper 80-42.2.

Electric Power Research Institute (EPRI). 1984. Continuous emission monitoring guidelines. Report No. EPRI CS-3723. EPRI, Palo Alto, CA.

Electric Power Research Institute. 1988. Continuous emission monitoring guidelines—update. Report No. EPRI CS-5998. EPRI, Palo Alto, CA.

Hamil, H. F. 1980. The development of performance audit devices for optical transmissometers. Unpublished (EPA contract 68-02-2489. EPA Quality Assurance Division, Research Triangle Park, NC 27711. Draft final report.)

International Standards Organization (ISO). 1981. *Gas Analysis Standards—Preparation of Calibration Gas Mixtures*. ISO 6142, ISO 6143, ISO 6144, ISO 6711. ISO, Geneva, Switzerland.

ISO. 1984. *Gas Analysis—Preparation of Calibration Gas Mixtures. Mass Dynamic Method*. ISO 7395. ISO, Geneva, Switzerland.

ISO. 1986. *Gas Analysis Standards—Preparation of Calibration Gas Mixtures*. ISO 6145/1, ISO 6145/3, ISO 6145/4. ISO, Geneva, Switzerland.

Jernigan, J. R. 1986. Assessment of the equivalency of three-run accuracy audits versus six-run relative accuracy audits for characterizing CEMS performance. In *Transactions—Continuous Emission Monitoring: Advances and Issues*. Air Pollution Control Association, Pittsburgh, pp. 332–342.

Kopecky, M. J., and Rodger, B. 1979. Quality assurance for procurement of air analyzers. In *ASQC Technical Conference Transactions—Houston, 1979*. pp. 35–40.

McCoy, P. G. 1990. The "CEM Nation" an analysis of U.S. EPA's Database—1988. In *Proceedings—Specialty Conference on Continuous Emission Monitoring: Present and Future Applications*. Air and Waste Management Association, Pittsburgh, pp. 10–36.

Noland, S., and Reynolds, W. E. 1990. Design and operation of a transportable emission monitoring system utilizing dilution probe technology. In *Proceedings—Specialty Conference on Continuous Emission Monitoring: Present and Future Applications*. Air and Waste Management Association, Pittsburgh, pp. 136–147.

Peeler, J. W. 1990. Guidelines for CEMS performance specifications and quality assurance requirements for municipal waste combustion facilities. NESCAUM, U.S. EPA Contract No. 68D90055. (Obtain by writing NESCAUM, 85 Merrimac St., Boston, MA 02114.)

Peeler, J. W., and Deaton, G. D. 1981. Field testing of a transportable extractive monitoring system for $SO_2$, $NO/NO_x$, CO and $CO_2$. In *Proceedings—Continuous Emission Monitoring: Design, Operation and Experience*. Air Pollution Control Association, Pittsburgh, pp. 176–186.

Peeler, J. W., Fox, V. L., and Plaisance, S. J. 1988. *Inspection Guide for Opacity Continuous Emission Monitoring Systems*. EPA 340/1-88-002.

Plaisance, S. J., and Peeler, J. W. 1987. *Technical Assistance Document: Performance Audit Procedures for Opacity Monitors*. EPA 600/8-87-025.

Reynolds, W. E. 1989. *Field Inspector's Audit Techniques: Gas CEMS's Which Accept Calibration Gases*. EPA 340/1-89-003.

U.S. Environmental Protection Agency (U.S. EPA). 1976 (1/9/84 update). *Quality Assurance Handbook for Air Pollution Measurement Systems*, Vol. 1, *Principles*. EPA 600/9-76-005.

U.S. EPA. 1977a. *Quality Assurance Handbook for Air Pollution Measurement Systems*, Vol. 3, *Stationary Source Specific Methods*. EPA 600/4-77-027b.

U.S. EPA. 1977b (6/15/78 update). Traceability protocol for establishing true concentration of gases used for calibration and audits of continuous source emission monitors (protocol no. 1). In *Quality Assurance Handbook for Air Pollution Measurement Systems*, Vol. 3, *Stationary Source Specific Methods*. EPA 600/4-77-027b. Section 3.0.4.

U.S. EPA. 1977c (6/1/86 update). Continuous emission monitoring (CEM) systems good operating practices. In *Quality Assurance Handbook for Air Pollution Measurement Systems*, Vol. 3, *Stationary Source Specific Methods*. EPA 600/4-77-027b. Section 3.0.9.

U.S. EPA. 1977d (6/9/87 update). Procedure for NBS-traceable certification of compressed gas working standards used for calibration and audit of continuous source emission monitors (revised traceability protocol no. 1). In *Quality Assurance Handbook for Air Pollution Measurement Systems*, Vol. 3, *Stationary Source Specific Methods*. EPA 600/4-77-027b. Section 3.0.4.

U.S. EPA. 1990a. Performance specifications. In *Code of Federal Regulations—Protection of the Environment*. 40 CFR 60 Appendix B.

U.S. EPA. 1990b. Reference methods. In *Code of Federal Regulations—Protection of the Environment*. 40 CFR 60 Appendix B.

U.S. EPA. 1990c. Quality assurance procedures. In *Code of Federal Regulations—Protection of the Environment*. 40 CFR 60 Appendix F.

**Bibliography**

American Society for Testing and Materials. 1975. *Calibration in Air Monitoring*. ASTM, Philadelphia.

Caron, A. L., and Hoy, D. R. 1987. A short-term evaluation of stability of a dynamic dilution system proposed to challenge TRS emission monitors on Kraft emission sources. In *Proceedings—Transactions of the Pulp and Paper Institute (TAPPI) 1987 Environmental Conference, Atlanta* (Paper 12-2). pp. 209–215 .

Decker, C. E., Saeger, M. L., Eaton, W. C., and von Lehmden, D. J. 1981. Analysis of commercial cylinder gases of nitric oxide, sulfur dioxide, and carbon monoxide at source concentrations. In *Proceedings—Specialty Conference on Continuous Emission Monitoring: Design, Operation and Experience*. Air Pollution Control Association, Pittsburgh, pp. 197–209.

Entropy Environmentalists, Inc. 1983. *Assessment of the Adequacy of the Appendix F Quality Assurance Procedures for Maintaining CEMs Data Accuracy: Status Report #1*. EPA 600/4-83-047.

Floyd, J. R. 1979. Quality assurance audit techniques for validating data from continuous emission stack monitoring (CEM) systems. Paper presented at the Air Pollution Control Association Meeting, Cincinnati. Paper 79-20.5.

Floyd, J. R. 1980. Application and results of quality assurance (QA) audits and other novel techniques to continuous emission monitor (CEM) programs. Paper presented at the Air Pollution Control Association Meeting, Montreal. Paper 80-70.3.

Host, J. T., Gilman, B. H., and McMillan, T. S. 1981. San Diego Gas & Electric's CEMQA program—a practical approach. In *Proceedings—Continuous Emission Monitoring: Design, Operation and Experience*. Air Pollution Control Association, Pittsburgh, pp. 210–220.

Hughes, E., and Mandel, J. 1981. *A Procedure for Establishing Traceability of Gas Mixtures to Certain National Bureau of Standards Standard Reference Materials*. EPA 600/7-81-010.

Hughes, E. E. 1981. Certified reference materials for continuous emission monitoring. In *Proceedings—Continuous Emission Monitoring: Design, Operation and Experience*. Air Pollution Control Association, Pittsburgh, pp. 187–196.

Hughes, E. E. 1982. Certified reference materials for continuous emission monitoring. *J. Air Pollut. Control Assoc.* 32:708–711.

Jahnke, J. A. 1982. Conference overview. In *Proceedings—Specialty Conference on Continuous Emission Monitoring: Design, Operation and Experience*. Air Pollution Control Association, Pittsburgh, pp. 308–311.

Jahnke, J. A. 1984. *Transmissometer Systems—Operation and Maintenance, An Advanced Course*. APTI Course 476A. EPA 450/2-84-004.

James, R. E., and Wolbach, C. D. 1976. Quality Assurance of Stationary Source Emission Monitoring Data. Institute of Electrical and Electronics Engineers, New York, Annals No. 75 CH1004-1 36-2.

James, R. E. 1979. Quality assurance of data from $SO_2$ and $NO_x$ monitors required by EPA new source performance standards. In *Proceedings—Specialty Conference on Quality Assurance in Air Pollution Measurement*. Air Pollution Control Association, Pittsburgh, pp. 419–429.

Logan, T. J., and Midgett, M. R. 1979. Quality assurance programs to support the use of continuous emission monitors for direct compliance. In *Proceedings—Specialty Conference on Quality Assurance in Air Pollution Measurement*. Air Pollution Control Association, Pittsburgh, pp. 413–418.

Logan, T. J., Rollins, R., and Jernigan, J. R. 1984. Quality assurance for compliance continuous emission monitoring systems: Evaluation of span drift for gas CEMS. In *Proceedings—Specialty Conference on Quality Assurance in Air Pollution Measurements*. Air Pollution Control Association and American Society for Quality Control, Pittsburgh, Conference Reprint. Session 108.

Nazzaro, J. C. 1986. Continuous emission monitoring system approval, auditing and data processing in the Commonwealth of Pennsylvania. In *Transactions—Continuous Emission Monitoring: Advances and Issues*. Air Pollution Control Association, Pittsburgh, pp. 175–186.

Nolen, S. L., Ryan, J. V., and Bridge, R. 1987. Real-time monitoring of a hazardous waste incinerator with a mobile laboratory. Paper presented at the Air Pollution Control Association Meeting, New York. Paper 87-23.5.

Oldaker, G. B., Rosenquest, J. M., and Purcell, R. Y. 1981. Quality assurance evaluation of transmissometers. In *Proceedings—Specialty Conference on Continuous Emission Monitoring: Design, Operation and Experience*. Air Pollution Control Association, Pittsburgh, pp. 285–292.

Peeler, J. W. 1986a. *Recommended Quality Assurance Procedures for Opacity Continuous Emission Monitoring Systems*. EPA 340/1-86-010.

Peeler, J. W. 1986b. *CEMS Pilot Project: Evaluation of Opacity CEMS Reliability and QA Procedures*, Vol. 1. EPA 340/1-86-009a.

Peeler, J. W. 1986c. *CEMS Pilot Project: Evaluation of Opacity CEMS Reliability and QA Procedures*, Vol. 2 (Appendices). EPA 340/1-86-009b.

Plaisance, S. J. 1988. Advances in opacity CEMS performance auditing. Paper presented at the Air Pollution Control Association Meeting, Dallas. Paper 88-137.7.

Purcell, R. Y., and Rosenquest, J. M. 1982. Field performance audit procedures for opacity monitors. Environmental Protection Agency CEM Report Series No. 5-271-7/82.

Reynolds, W. E. 1984. Development and evaluation of SO$_2$ CEM QA procedures. In *Quality Assurance in Air Pollution Measurements*. Air Pollution Control Association and American Society for Quality Control, Pittsburgh, pp. 752–760.

Rollins, R., Logan, T. J., and Midgett, M. R. 1985. An assessment of the long-term precision of continuous emission monitors. Paper presented at the Air Pollution Control Association Meeting, Detroit. Paper 85-51.2.

U.S. Environmental Protection Agency (U.S. EPA). 1977a (7/1/79 update). *Quality Assurance Handbook for Air Pollution Measurement Systems*, Vol. 2, *Ambient Air Specific Methods*. EPA 600/4-77-027a.

U.S. EPA. 1977b (11/5/85 update). Calculations and interpretation of accuracy for continuous emission monitoring systems (CEMS). In *Quality Assurance Handbook for Air Pollution Measurement Systems*, Vol. 3, *Stationary Source Specific Methods*. EPA 600/4-77-927a. Section 3.0.7.

U.S. EPA. 1977c (11/26/85 update). Guideline for developing quality control procedures for gaseous continuous emission monitoring systems. In *Quality Assurance Handbook for Air Pollution Measurement Systems*, Vol. 3, *Stationary Source Specific Methods*. EPA 600/4-77-027b. Section 3.0.10.

U.S. EPA. 1982. Performance audit procedures for SO$_2$, NO$_x$, CO$_2$, and O$_2$ continuous emission monitors. CEM Report Series No. 5-371-7/82. EPA Division of Stationary Source Enforcement.

U.S. EPA. 1983. *Performance Audit Procedures for Opacity Monitors*. EPA 340/1-83-010.

U.S. EPA. 1990. Test methods. In *Code of Federal Regulations—Protection of the Environment*. 40 CFR 60 Appendix A.

U.S. EPA—Office of Air Quality Planning and Standards. 1980. *Interim Guidelines and Specifications for Preparing Quality Assurance Project Plans*. QAMS-005/80.

Van Gieson, J., and Paley, L. R. 1984. *Summary of Opacity and Gas CEMS Audit Programs*. EPA 340/1-84-016.

von Lehmden, D. J., and Walsh, G. W. 1990. Appendix F DARs for CEMS at Subpart D facilities. In *Proceedings—Specialty Conference on Continuous Emission Monitoring: Present and Future Applications*. Air and Waste Management Association, Pittsburgh, pp. 103–119.

Walsh, G., 1989. *Data Assessment Reports for CEMS at Subpart Da Facilities*. EPA 600/3-89-027.

Wohlschlegal, P. 1976. *Guidelines for Development of a Quality Assurance Program*, Vol. 15, *Determination of Sulfur Dioxide Emissions from Stationary Sources by Continuous Monitors*. EPA 650/4-74-005o.

Wright, R. S., Decker, C. E., and Barnard, W. F. 1986. Performance audit of inspection and maintenance calibration gases. Paper presented at the Air Pollution Control Association Meeting, Minneapolis. Paper 86-46.4.

Wright, R. S., Eaton, W. C., and Decker, C. E. 1987. *NBS/EPA Certified Reference Material Performance Audit Program: Status Report 2*. EPA 600/S4-86/045.

Wright, R. S., Tew, E. L., Decker, and von Lehmden, D. J. 1986. Analysis of EPA protocol gases used for calibration and audits of continuous emission monitoring systems and ambient air analyzers—results for audit 6. In *Transactions —Continuous Emission Monitoring: Advances and Issues*. Air Pollution control Association, Pittsburgh, pp. 343–355.

Wright, R. S., Tew, E. L., Decker, C. E., von Lehmden, D. J., and Barnard, W. F. 1987. Performance audits of EPA protocol gases and inspection and maintenance calibration gases. *J. Air & Waste Mgmt. Assoc.* 37:384–385.

Wright, R. S., Wall, C. V., Decker, C. E., and von Lehmden, D. J. 1989. Accuracy assessment of EPA protocol gases in 1988. *J. Air & Waste Mgmt. Assoc.* 39:1225–1227.

# A

# *F* Factors

*F* factors are used for calculating flue-gas emissions in terms of the emission rate *E*, expressed in nanograms per joule [or pounds per million British thermal units (Btu)]—units of the standard frequently used for combustion sources. This process rate standard is an expression of the mass or weight of emissions emitted per unit of energy generated by the source.

Using the pollutant mass ($pmr_s$) discussed in Chapter 6, the emission rate *E* is given as

$$E = \frac{pmr_s}{Q_H} = \frac{c_s Q_s}{Q_H} = \frac{c_s A_s V_s}{R_f \, GCV} \tag{A-1}$$

where $pmr_s$ = mass emission rate (in grams per hour or pounds per hour)

$c_s$ = pollutant concentration [in grams per dry standard cubic meter (dscm) or pounds per dry standard cubic foot (dscf)]

$Q_s$ = flue-gas volumetric flow rate (in dscm per hour or dscf per hour)

$A_s$ = stack or duct area at the velocity measurement cross section (in square meters or square feet)

$v_s$ = flue-gas velocity (in meters per hour or feet per hour)

$Q_H$ = combustion source heat input rate (in joules per hour or Btu per hour)

$R_f$ = fuel feed rate (in kilograms per hour or pounds per hour)

GCV = gross calorific value (high heating value) of the fuel (in joules per kilogram or Btu per pound) (the total heat obtained from the complete combustion of a fuel, referenced to a set of standard conditions)

285

For CEM systems, the determination of emissions by using Equation (A-1) is cumbersome because measurements for the pollutant concentration, flue-gas velocity, fuel feed rate, and fuel heating value must be made on a continual basis. Errors associated with each of these parameters can combine to give a result that may be neither sufficiently accurate nor precise to serve regulatory or technical purposes. Fortunately, alternate methods have been developed (Shigehara, Neulicht, and Smith 1973; Neulicht 1975) that simplify the calculation by considering the known stochiometry associated with combusted fuels. These methods, the so-called $F$ factor methods, can be expressed in the general form of Equation (A-2):

$$E = c_s F D_{corr} \qquad \text{(A-2)}$$

where $F$ = an $F$ factor (in dscm per joule or dscf per cubic foot)
$D_{corr}$ = a dilution correction factor (dimensionless)

The expression is obviously dimensionally correct because (g/dscm) × (dscm/J) gives the process emission rate in grams per joule (g/J). An $F$ factor is merely the ratio of the theoretical volume of gas generated (by the complete combustion of a quantity of fuel) to the amount of heat produced by the fuel upon combustion. $F$ factors are calculated from a consideration of the theoretical chemical equations of combustion and the laboratory determination of the gross calorific value of the fuel. The utility of $F$ factors lies in the fact that their values are very consistent for a given fuel category. Once determined for bituminous coal, oil, or natural gas, they can be assumed for that fuel category without recalculation. There are several types of $F$ factor: the $F_d$ or dry, $F$ factor; the $F_c$ factor; and the $F_w$, or wet, $F$ factor. Each of these will be discussed in this appendix.

## THE OXYGEN-BASED DRY $F$ FACTOR, $F_d$

The dry $F$ factor ($F_d$) is the ratio of the theoretical volume of dry gases ($V_t$) given off by complete combustion of a known amount of fuel to the high heating value of the burned fuel:

$$F_d = \frac{\text{volume dry combustion gases per kilogram}}{\text{gross calorific value per kilogram}} = \frac{V_t}{GCV} \qquad \text{(A-3)}$$

The values of the constituents in the $F$ factor are determined by fuel analysis. There are two types of fuel analysis, proximate and ultimate analysis:

**Proximate analysis**   a fuel analysis procedure that expresses the principal characteristics of the fuel as follows: (1) percentage of moisture; (2) percentage of ash; (3) percentage of volatile matter; (4) percentage of fixed carbon; (5) percentage of sulfur; (6) heating value; (7) ash fusion temperature

**Ultimate analysis**   the determination of the exact chemical composition of the fuel without paying attention to the physical form in which the compounds appear. The analysis is generally given in terms of percentage hydrogen, percentage carbon, percentage sulfur, percentage nitrogen, and percentage oxygen.

The data generated in an ultimate analysis of a given fuel allow the calculation of an $F_d$ factor based on the composition of the fuel constituents. The individual chemical components are included in the theoretical volume as uncombined elements. Each contributes to the total volume $V_t$ based upon the percentage present in the fuel. An $F$ factor can then be calculated for any fuel when the percentage composition of each constituent is known:

$$F_d = \frac{10^{-5}[22.7(\%H) + 9.57(\%C) + 3.54(\%S) + 0.86(\%N) - 2.85(\%O)]}{GCV} \qquad \text{metric units} \quad \text{(A-4)}$$

$$F_d = \frac{10^{6}[3.64(\%H) + 1.53(\%C) + 0.57(\%S) + 0.14(\%N) - 0.46(\%O)]}{GCV} \qquad \text{English units} \quad \text{(A-5)}$$

*Note:* Units for the conversion factors in the expressions are $10^{-5}$ kJ/J and $10^6$ Btu/million Btu for GCV expressed in kilojoules per kilogram and in Btu per pound, respectively. The constants in the expressions are given in units of standard cubic meters per kilogram (e.g., 22.7 scm/kg) and standard cubic feet per pound (e.g., 3.64 scf/lb).

The preceding equations account for only a stoichiometric amount of oxygen—that amount of oxygen necessary to oxidize the fuel completely to its combustion products. An industrial facility burning large quantities of fuel adds a stoichiometric amount of air (oxygen and nitrogen) and some excess air to assure complete combustion of the fuel. The volume of the combustion products is related to the excess air as follows:

$$V_t = Q_s \times [\text{dilution air correction term}] \qquad \text{(A-6)}$$

The stoichiometric oxygen present would be consumed for combustion of the fuel. The remaining oxygen present in the combustion gases is, therefore, an excess amount and dilutes the volume of gas generated by combustion. As a result, the value of $Q_s$ would be higher, the more excess air is used and must be corrected in order to calculate $V_t$. For dry air containing 20.9% oxygen, this correction term is

$$[\text{dilution air correction term}] = \frac{20.9 - \%O_{2\,d}}{20.9} \qquad \text{(A-7)}$$

Then

$$V_t = Q_s \times \frac{20.9 - \%O_{2\,d}}{20.9} \qquad \text{(A-8)}$$

Because the heat released by the fuel is not affected by the dilution air, Equation (A-3) becomes

$$F_d = \frac{V_t}{\text{HHV}_f} = \frac{V_t}{Q_H} = \frac{Q_s}{Q_H}\left[\frac{20.9 - \%O_{2\,d}}{20.9}\right] \qquad \text{(A-9a)}$$

or

$$\frac{Q_s}{Q_H} = F_d\left[\frac{20.9}{20.9 - \%O_{2\,d}}\right] \qquad \text{(A-9b)}$$

Therefore, from Equation (A-1), with $c_d$ equal to the pollutant concentration measured on a dry basis,

$$E = c_d F_d\left[\frac{20.9}{20.9 - \%O_{2\,d}}\right] \qquad \text{(A-10)}$$

The dry $F$ factor is used for emission rate calculations when both the pollutant gas and oxygen concentrations are measured by the CEM system on a *dry* basis. If the moisture content of the flue gas ($B_{ws}$) is measured, known, or assumed, the dry $F$ factor can be used by modifying Equation (A-10):

$$E = c_w F_d \left[ \frac{20.9}{20.9(1 - B_{ws}) - \%O_{2w}} \right] \qquad \text{(A-11)}$$

where $c_w$ = pollutant concentration measured on a wet basis
$\%O_{2w}$ = oxygen percentage measured on a wet basis

## THE OXYGEN-BASED WET $F$ FACTOR, $F_w$

In-situ and dilution CEM systems measure flue gases on a wet basis. If oxygen is measured for the dilution correction, the flue-gas moisture content must be known to determine $E$, when using $F_d$. For accurate measurements in sources where moisture content fluctuates, a moisture monitor or monitoring system is therefore necessary. This is an added expense and an additional source of error in the emission rate calculation. Under certain conditions, however, another $F$ factor—the wet $F$ factor—can be used (Shigehara and Neulicht 1976).

The wet $F$ factor $F_w$ is defined as the ratio of the quantity of *wet* effluent gas generated by combustion to the gross calorific value ($\text{GCV}_w$) of the fuel determined for the fuel on an "as received" or "as fired" basis. $\text{GCV}_w$ includes the free water in the fuel.

$$F_w = \frac{10^{-5}[34.74(\%H) + 9.57(\%C) + 3.54(\%S) + 0.86(\%N) - 2.85(\%O) + 1.30(\%H_2O)]}{\text{GCV}_w}$$

metric units  (A-12)

$$F_w = \frac{10^{6}[5.57(\%H) + 1.53(\%C) + 0.57(\%S) + 0.14(\%N) - 0.46(\%O) + 0.21(\%H_2O)]}{\text{GCV}_w}$$

English units  (A-13)

The $H_2O$ percentage can be omitted if the oxygen and hydrogen from the water is included in the values for $\%H$ and $\%O_2$.

Because excess air used for combustion may include ambient moisture, an ambient moisture fraction $B_{wa}$, must be included in the emission rate calculation. The expression is

$$E = c_w F_w \left[ \frac{20.9}{20.9(1 - B_{wa}) - \%O_{2w}} \right] \tag{A-14}$$

Any of the following values may be used for $B_{wa}$ in this expression:

1. fixed constant value of 0.027
2. continuous measured value
3. monthly value based on previous history
4. annual value based on previous history

Although the value for $B_{wa}$ may be approximate, its contribution to the overall expression is not that great. The estimates should not introduce a negative error greater than $-1.5\%$. However, positive errors (overestimation of emissions) of as much as 5% have been noted in some geographic locations (U.S. EPA 1991). The expression cannot be used in any process in which water is introduced into *or* removed from the flue gas stream. It is therefore not applicable to CEM systems installed after wet scrubbers.

## THE CARBON DIOXIDE–BASED $F$ FACTOR, $F_c$

If a carbon dioxide analyzer is used in the CEM system, the carbon dioxide $F$ factor ($F_c$) can be used. $F_c$ is the ratio of the theoretical volume of carbon dioxide produced during the combustion of a given amount of fuel, to its gross calorific value:

$$F_c = \frac{20.0(\%C)}{GCV} \quad \text{metric units} \tag{A-15}$$

$$F_c = \frac{0.321(\%C)}{GCV} \quad \text{English units} \tag{A-16}$$

The corresponding emission rate using the $F_c$ factor is given in Equations (A-17a) and (A-17b):

$$E = c_d F_c \frac{100}{\%CO_{2d}} \tag{A-17a}$$

$$E = c_w F_c \frac{100}{\%CO_{2w}} \tag{A-17b}$$

The $F_c$ factor can be used on either a wet basis or a dry basis, provided that both the pollutant gas concentration and the $CO_2$ percentage are measured on the same basis. That is, if $c_d$ and $\%CO_{2\,d}$ or $c_w$ and $\%CO_{2\,w}$ are used. $F_c$ factor calculations are therefore very useful in in-situ path CEM systems, dilution systems, or hot–wet extraction systems that incorporate a $CO_2$ analyzer.

When the pollutant concentration is measured on a wet basis and the $CO_2$ percentage is measured on a dry basis, the following equation can be used:

$$E = \frac{c_w F_c}{(1 - B_{ws})} \frac{100}{\%CO_{2\,d}} \tag{A-18}$$

If the reverse is true, the following expression may be used:

$$E = c_d(1 - B_{ws}) F_c \frac{F_c 100}{\%CO_{2\,w}} \tag{A-19}$$

## TABULATED *F* FACTORS AND CONVERSION FACTORS

Because most pollutant analyzers express emission concentration in terms of parts per million (ppm), the analyzer values must first be converted into units of nanograms per standard cubic meter (ng/scm) or pounds per standard cubic foot (lb/scf) before substitution into the *F* factor expressions. Typical conversions are

$$1 \text{ ng/scm } SO_2 = 2.66 \times 10^6 \text{ ppm } SO_2$$

$$1 \text{ lb/scf } SO_2 = 1.66 \times 10^{-7} \text{ ppm } SO_2$$

$$1 \text{ ng/scm } NO_x = 1.912 \times 10^6 \text{ ppm } NO_x$$

$$\text{lb/scf } NO_x = 1.194 \times 10^{-7} \text{ ppm } NO_x$$

$$1 \text{ ng/scm} = 1.602 \times 10^{13} \text{ lb/scf}$$

The *F* factors have been calculated and tabulated by the U.S. EPA using data obtained from the literature and have been found to be quite consistent within a fuel category. For example, the maximum percentage deviation from the midpoint $F_d$ factor for bituminous coal was found to be 3.1% (Shigehara et al. 1978). The *F* factors may be determined by the plant using the appropriate equations given previously; however they

**TABLE A-1**    **$F$ Factors for Various Fuels (U.S. EPA 1991)[a]**

| Fuel Type | $F_d$ | | $F_w$ | | $F_c$ | |
|---|---|---|---|---|---|---|
| | dscm/J | dscf/(10$^6$ Btu) | wscm/J | wscf/(10$^6$ Btu) | scm/J | scf/(10$^6$ Btu) |
| Coal | | | | | | |
|   Anthracite[b] | $2.71 \times 10^{-7}$ | 10,100 | $2.83 \times 10^{-7}$ | 10,540 | $0.530 \times 10^{-7}$ | 1,970 |
|   Bituminous[b] | $2.63 \times 10^{-7}$ | 9,780 | $2.86 \times 10^{-7}$ | 10,640 | $0.484 \times 10^{-7}$ | 1,800 |
|   Lignite | $2.65 \times 10^{-7}$ | 9,860 | $3.21 \times 10^{-7}$ | 11,950 | $0.513 \times 10^{-7}$ | 1,910 |
| Oil[c] | $2.47 \times 10^{-7}$ | 9,190 | $2.77 \times 10^{-7}$ | 10,320 | $0.383 \times 10^{-7}$ | 1,420 |
| Gas | | | | | | |
|   Natural | $2.43 \times 10^{-7}$ | 8,710 | $2.85 \times 10^{-7}$ | 10,610 | $0.287 \times 10^{-7}$ | 1,040 |
|   Propane | $2.34 \times 10^{-7}$ | 8,710 | $2.74 \times 10^{-7}$ | 10,200 | $0.321 \times 10^{-7}$ | 1,190 |
|   Butane | $2.34 \times 10^{-7}$ | 8,710 | $2.79 \times 10^{-7}$ | 10,390 | $0.337 \times 10^{-7}$ | 1,250 |
| Wood | $2.48 \times 10^{-7}$ | 9,240 | | | $0.492 \times 10^{-7}$ | 1,830 |
| Wood bark | $2.58 \times 10^{-7}$ | 9,600 | | | $0.516 \times 10^{-7}$ | 1,920 |
| Municipal solid waste | $2.57 \times 10^{-7}$ | 9,570 | | | $0.488 \times 10^{-7}$ | 1,820 |

[a] Determined at U.S. EPA standard conditions: 20°C (68°F) and 760 mmHg (299.92 in. Hg).
[b] As classified according to ASTM D388-77
[c] Crude, residual, or distillate

should lie within 2–5% of the EPA tabulated values [see Shigehara (1978) for maximum deviations]. The EPA tabulated values are given in Table A-1.

$F_w$ factors for wood, wood bark, and municipal solid waste are not tabulated, due to the high variability of free moisture content in the materials. Because the composition of municipal solid waste is also highly variable, some care should be taken in the application of $F$ factors for the waste fuel.

For sources using a combination of fossil fuels, a combination $F$ factor may be determined from the general expression

$$F_m = \sum_{i=1}^{n} x_i F_i \qquad \text{(A-20)}$$

where $F_m$ = the $F$ factor for the fuel mixture
$\quad\quad\, x_i$ = the fraction of total heat input from each type of fuel
$\quad\quad\, F_i$ = the appropriate $F$ factor for each type of fuel ($F_d$, $F_w$, or $F_c$)
$\quad\quad\, n$ = number of fuels being burned in combination

### Other Uses of F factors

$F$ factors provide a convenient means of performing certain checks and cross-checks of combustion and emissions data. For example, if both oxygen and carbon dioxide are measured on a dry basis, another factor, the $F_o$ factor, can be derived from Equations (A-10) and (A-17a):

$$F_o = \frac{20.9}{100} \frac{F_d}{F_c} = \frac{20.9 - \%O_{2\,d}}{\%CO_{2\,d}} \qquad \text{(A-21)}$$

Because $F_d$ and $F_c$ are relatively constant for a given fuel category, the $F_o$ factor will also be as constant. The substitution of $O_2$ and $CO_2$ data into the right-hand side of the equation should then give a value equal to the factor. The expression is useful in checking Orsat analyses as well as CEM data.

If the flue-gas volumetric flow rate $Q_s$ and the heat input rate $Q_H$ are known, the $F_d$ factor can be determined:

$$F_d = \frac{Q_s}{Q_H} \frac{(20.9 - \%O_{2\,d})}{20.9} \qquad \text{(A-22)}$$

Alternatively, if $Q_H$ is known, by using the tabulated value for $F_d$ and the measured value for $\%O_{2\,d}$ Equation (A-22) can be manipulated to obtain the flue-gas volumetric flow rate. This is basically a material balance check that is consistent with the combustion stoichiometry. Similarly, if $Q_s$ is known, then the heat input rate can be calculated.

Similar expressions can be used if a wet basis oxygen analyzer or a carbon dioxide analyzer is used in the CEM system [Equations (A-23)–(A-25)]:

$$F_d = \frac{Q_s}{Q_H} \frac{20.9(1 - B_{wa}) - \%O_{2w}}{20.9} \tag{A-23}$$

$$F_c = \frac{Q_s}{Q_H} \frac{\%CO_{2d}}{100} \tag{A-24}$$

$$F_c = \frac{Q_{sw}}{Q_H} \frac{\%CO_{2w}}{100} \tag{A-25}$$

The wet and dry $F$ factors can be manipulated to obtain the flue-gas moisture content $B_{ws}$, if the oxygen is measured on a wet basis (McGowan 1976). This is an interesting derivation because it allows the application of in-situ analyzers after wet scrubber systems without requiring the installation of a moisture monitor. This expression is given in Equation (A-26):

$$B_{ws} = 1 - \frac{F_d}{F_w}(1 - B_{wa}) - \frac{\%O_{2w}}{20.9}\left(1 - \frac{F_d}{F_w}\right) \tag{A-26}$$

A similar expression can be derived if wet basis oxygen and carbon dioxide measurements are made by the CEM system (Aldina 1985).

$$B_{ws} = 1 - \frac{O_{2w} + CO_{2w}F_o}{20.9} \tag{A-27}$$

Also, if the emission rate $E$ is known by some independent means, the moisture content can be determined using Equation (A-28):

$$B_{ws} = 1 - \frac{O_{2w}}{20.9} \frac{c_w F_d}{E} \tag{A-28}$$

It should be noted that there are a number of variables in Equations (A-26)–(A-28), each having its own accuracy and precision. The adequacy of using this technique in any given application should certainly be cross-checked using an appropriate direct measurement for $B_{ws}$.

## Errors and Problems in the Use of *F* Factors

Tabulated *F* factors do have a certain degree of imprecision because the values only represent midpoint values from a finite set of information. In general, however, maximum deviations from the midpoint value should be on the order of 3%.

The *F* factors assume complete combustion of the fuel. Incomplete combustion can cause an error, but this can be corrected if CO concentrations are determined. Corrections are performed by adjusting the $O_2$ or $CO_2$ percentages as follows:

$$(\%O_2)_{adj} = \%O_2 - 0.5(\%CO) \qquad \text{(A-29)}$$

$$(\%CO_2)_{adj} = \%CO_2 + \%CO \qquad \text{(A-30)}$$

In most cases, however, CO levels are on the order of a few 100 ppm and do give a significant adjustment.

### References

Aldina, G. J. 1985. Continuous emissions monitoring system for dry basis pollutant mass rate measurements. Private communication.

McGowan, G. F. 1976. Alternative methods for the reduction of dry and wet based measurement data. Lear Siegler, Inc., Technical Communication.

Neulicht, R. M. 1975. Emission correction factor for fossil fuel–fired steam generators: $CO_2$ concentration approach. *Stack Sampling News* 2(8): 6–11.

Shigehara, R. T., and Neulicht, R. M. 1976. Derivation of Equations for calculating power plant emission rates, $O_2$ based method—wet and dry measurements. Emission Measurement Branch, ESED, OAQPS, U.S. Environmental Protection Agency.

Shigehara, R. T., Neulicht, R. M., and Smith, W. S. 1973. A method for calculating power plant emissions. *Stack Sampling News* 1(1): 5–9.

Shigehara, R. T., Neulicht, R. M., Smith, W. S., and Peeler, J. W. 1978. Summary of *F* factor methods for determining emissions from combustion sources. In *Stack Sampling Technical Information—A Collection of Monographs and Papers*. Vol. II EPA-450/2-78-0426, pp. 29–43.

U.S. Environmental Protection Agency (U.S. EPA) 1991. Method 19—determination of sulfur dioxide removal efficiency and particulate matter, sulfur dioxide, and nitrogen oxides emission rates. In *U.S. Code of Federal Regulations*. 40 CFR 60 Appendix A.

# B

## Conversion Factors and Useful Information

### Constants

| | |
|---|---|
| Avogadro's number | $6.02 \times 10^{23}$ atoms per gram-atom |
| Faraday constant | $9.65 \times 10^4$ Coulombs per mole |
| Gas constants | $82.05$ atm $\cdot$ cm$^3$/(g-mol $\cdot$ K) |
| | $1.987$ cal/(g-mol $\cdot$ K) |
| | $10.731$ ft $\cdot$ lb $\cdot$ in.$^2$/(lb-mol $\cdot$ R) |
| | $0.732$ ft$^3$ $\cdot$ atm/(lb-mol $\cdot$ R) |
| 1 g-mol | $22.4$ l ideal gas at standard temperature and pressure (STP) |
| 1 lb-mol | $359$ ft$^3$ ideal gas at STP |
| $\ln 10$ | $2.3026$ |
| Natural log base $e$ | $2.7183$ |
| Planck's constant | $6.62 \times 10^{-27}$ erg-s |
| Speed of light | $3.0 \times 10^{10}$ cm/s |
| Speed of sound | $344$ m/s in air at $20°C$ |

### EPA STANDARD CONDITIONS

$T_{std} = 20°C \ (68°F)$
$P_{std} = 29.92$ in. Hg (1 atm)

## CONVERSION EXPRESSIONS

### Temperature

$K = °C + 273.16$    Degrees kelvin
$R = °F + 459.4$    Degrees Rankine
$°C = \frac{5}{9} (°F - 32)$
$°F = \frac{9}{5} (°C) + 32$

### Gas Concentration Units

To convert parts per million (ppm) to milligrams per cubic meter ($mg/m^3$) at a set of standard conditions:

$$\frac{mg}{dscm} = \frac{ppm \times MW}{22.414 \times (T_{std}/273.16)}$$

At EPA standard conditions:

100 ppm CO = 116 $mg/m^3$
100 ppm HCl = 163 $mg/m^3$
100 ppm $NO_2$ = 191 $mg/m^3$
100 ppm $SO_2$ = 266 $mg/m^3$

## CONVERSION FACTORS

### Energy

1 Btu = 1055 J

### Length

1 in. = 2.54 cm
1 ft = 0.305 m

### Mass

1 g = 0.0022 lb
1 lb = 453.6 g

### Mass Per Unit Volume

$1 \ g/m^3 = 0.0283 \ g/ft^3$
$1 \ lb/ft^3 = 16.02 \ kg/m^3$

298     Continuous Emission Monitoring

## Pressure

1 atm = $1.01325 \times 10^5$ Pa = 14.696 lb/in.$^2$
      = 760 torr = 407.2 in. $H_2O$

## Power

1 Btu/h = 0.2931 kW
1 kW = 3413 Btu/h
1 MW = 341,3000 Btu/h

## Volume

1 ft$^3$ = 0.0283 m$^3$ = 28.32 l
1 m$^3$ = 35.31 ft$^3$

# Index

299